TROUT AND SALMON CULTURE

(Hatchery Methods)

California Fish Bulletin Number 164

By

Earl Leitritz

Inland Fisheries Branch

and

Robert C. Lewis

Fishery Consultant

California Department of Fish and Game

1980

This printing is made possible by collaboration between the California Department of Fish and Game and the University of California Sea Grant Marine Advisory Program and the Division of Agricultural Sciences.

Originally published in 1959 as *Fish Bulletin No. 107* by the State of California, Department of Fish and Game. Revised in 1976 as *Fish Bulletin 164*.

Printed in the United States of America.

International Standard Book Number: 0-931876-36-2

TABLE OF CONTENTS

FOREWORD

This volume * has been prepared at the request of many of the Department's fish hatchery personnel. A hatchery treatise has long been needed to acquaint the beginning employee with the rudiments of fish culture, and also to act as a handy reference for those already experienced in the work. In addition, it should lead to greater uniformity in operations and to increased hatchery efficiency. It will also be helpful to the growing number of private trout hatchery operators.

Even though the art of trout culture dates back to the year 1741, when Stephen Ludwig Jacobi started artificial propagation in Germany, advances in methods and techniques were slow until shortly before World War II. During the past 10 or 12 years, applied science and mechanics have revolutionized fish hatchery operations. More advances have probably occurred during this period than since the very beginning of trout culture. The uses of new chemicals in treating diseases in hatcheries, eradicating undesirable fish populations, spawning, and transporting fish, and the employment of labor-saving devices such as fish loaders, self-graders, incubators, and dry feeds are only a few of the advances illustrating the progress made. They indicate that fish culture is at last beginning to receive the recognition and research that it deserves. With a greater demand for hatchery-reared fish each year, additional important advances are sure to take place.

In considering literature to be embodied in this volume and suggestions received from Department employees, attention was directed especially to subjects which would benefit the average hatcheryman and assist him with his everyday problems.

Earl Leitritz**
May, 1959

Seventeen years ago Earl Leitritz called attention to the rapid improvement in fish culture equipment and methods in the previous 10 or 12 years and correctly predicted continuing progress. California is constructing new hatcheries and rebuilding old ones with concrete ponds designed to operate with mechanical crowders and graders, fish pumps to grade and load fish, improved planting equipment, mechanical feeders, dry feed storage, recirculated hatchery water systems, mechanical egg sorters, mechanical equipment for taking spawn, and air spawning. An intensive selective breeding program has been initiated, methods developed for delayed fertilization of eggs, and many new operating techniques have been put into practice.

Great progress has been made in the past 17 years and many advances to this growing program will be made in the future. It is hoped that this bulletin will be updated again in another decade as it has been a "bible" to hatchery personnel and those closely related to it.

ROBERT C. "BOB" LEWIS
May 1976

* From California Department of Fish and Game, Fish Bulletin 107 (now out of print).
** Deceased March 2, 1968.

ACKNOWLEDGMENTS

The author wishes to acknowledge the encouragement given by Alex Calhoun in the preparation and arrangement of this volume.* It was through his efforts that its publication was made possible.

The experience and writings of many workers in the field of fish culture have been drawn on freely in compiling this publication. Special thanks are extended to those who contributed separate sections, in whole or in part, as follows: Robert C. Lewis, selective breeding of trout; Carl G. Hill and Lloyd C. Hume, incubator hatching; Howard L. Huddle, aquatic weed control; Robert Macklin, fish nutrition and feeding; Harold H. Hewitt, grading fish; Harold Wolf, fish diseases and treatment, and anesthetics and their use, Leo Shapovalov, classification of fishes.

Thanks are due the several hatchery managers who assisted in obtaining photographs and made facilities and material available during preparation.

Credit is due Cliffa E. Corson, who did the lettering and illustrations, and to Viola Kobriger, who provided valuable help in the final preparation of the copy.

Finally, thanks are due to Joseph H. Wales, who critically read the manuscript, and to Leo Shapovalov, who did the final editing.

EARL LEITRITZ
May, 1959

The author acknowledges the help given by many Fish and Game employees in supplying material for the revision and updating of this publication.

Special thanks are extended to Harold Wolf and William Wingfield in revising the sections on fish diseases and treatment, and anesthetics and their use; to George Bruley who obtained photographs of new equipment and facilities; to William Schafer who provided much valuable information and help with the manuscript; and to many hatchery managers and supervisors, particularly Andrew Weaver, who assisted with information and material.

ROBERT C. "BOB" LEWIS
May 1976

---

* From California Department of Fish and Game, Fish Bulletin 107 (now out of print).

Trout and Salmon Culture
(Hatchery Methods)

HATCHERY WATER SUPPLY

The characteristics of the watershed in which a hatchery is located, such as mineral content of rocks and soil, rainfall, hydraulic gradient, range of temperatures, and amount of foliage, control the primary characteristics of a hatchery water supply. Very heavy rainfall areas, headwater streams, and mountain locations with steep stream gradients are generally low and sometimes deficient in mineral content. Limestone areas supply calcium and magnesium, which are beneficial to growth and bone structure of fish. These waters also have a higher bicarbonate alkalinity, which tends to buffer and resist the effect of contaminating substances, such as acids or alkalies.

Moderately steep gradients above the intake are desirable for aeration; however, very steep and narrow canyons are most subject to flood damage. Good plant cover, such as trees, grass, and brush, is desirable because it minimizes erosion and silting of the hatchery water supply. Shade further tends to prevent the reflection of extremes of air temperature in changing water temperatures. A moderate and even temperature between 45 and 60° F, depending on the objectives of the installation, is desirable.

Sluggish streams, swamps, bogs, springs, and wells are often deficient in dissolved oxygen, low in pH, and high in free carbon dioxide. These disadvantages can usually be remedied by proper aeration. Springs and wells ordinarily have the advantages of moderate and uniform temperatures and absence of diseases and pollution.

Any human activity on a watershed above the hatchery supply, such as dairying, stock raising, agricultural activity, summer resorts, lumbering or mining, increases the tendency towards silting, organic or inorganic pollution, and changing stream stages and temperatures, all of which are reflected in a less desirable water supply for hatchery purposes.

A common difficulty in well supplies is that they are apt to have an excess of gas. This is sometimes indicated by a milky appearance as the water enters the troughs or ponds. It may cause severe fish losses due to the formation of gas bubbles in the blood, under the skin, and in the eyes. Adequate aeration before use in the hatchery will overcome this difficulty.

An excessively high mineral content, as at Fillmore Hatchery in southern California, may prevent normal development of trout eggs. On the other hand, such water may produce outstandingly rapid growth of fish after they are hatched.

Water from a large reservoir, properly aerated, can be a good supply. Multiple level outlets help correct water temperature problems by providing means to select cooler, lower level water during stratification. San Joaquin Hatchery in Fresno County is located below Friant Dam on the San Joaquin River and receives water from Millerton Lake, a fluctuating reservoir. Cool water is drawn from a low-level outlet and warmer water from a midlevel outlet. The water from lower levels of a reservoir must be aerated to dissipate undesirable gases and add oxygen (Figures 1 and 2).

FIGURE 1—Aerator at San Joaquin Hatchery, Fresno County, California. The water enters the aerator at the top and falls through a series of baffles and perforated plates to dissipate undesirable gases and add oxygen to the water. *Photograph by George Bruley, 1973.*

From the above discussion, it is evident that many factors should be considered in selecting and installing a satisfactory water supply. Desirable features may be summarized as follows: (1) moderate rainfall, (2) moderate gradient, (3) good cover, such as trees, grass, and brush, (4) adequate limestone and other mineral deposits, (5) uniform and moderate temperature, (6) freedom from grazing, logging, mining, and similar activities on the watershed above the hatchery supply, (7) a submerged intake, (8) a covered pipeline to minimize temperature changes, (9) a moderate gradient from intake to hatchery, (10) adequate aeration, and (11) enclosure and covering of the water supply to prevent surface contamination. A water supply from suitable springs is more desirable than one from wells, streams, or lakes.

FIGURE 2—A doughnut type aerator at Trinity River Hatchery, Lewiston, Trinity County, California. The water is sprayed from a circular pipe and falls into a center catch basin to dissipate undesirable gases and add oxygen to the water. *Photograph by George Bruley, 1973.*

Chemical Characteristics

Dissolved Oxygen

M. M. Ellis et al. (1946), in their studies on favorable water environment for freshwater fishes, conducted some 5,800 oxygen determinations at about 1,000 stations throughout the country. The analyses for dissolved oxygen on 1,300 samples from waters with a good-sized fish fauna ranged from 4 to 12 ppm, but 84% of the good faunas were found where oxygen values ranged from 5 to 9. They stated that 3 ppm is approximately the lethal point for freshwater fishes at summer temperatures but that respiratory difficulties developed below 5 ppm. The lowest safe level for trout is about 5 ppm and 7 ppm is preferable. One investigator referred to by Ellis states that 10 to 11 ppm is best for trout and that they may show discomfort when dissolved oxygen is less than 7.8.

The importance of adequate dissolved oxygen for fish warrants the inclusion of basic data on this subject.

Solubility of Oxygen

The solubility of oxygen for fresh water in equilibrium with air changes with altitude (Table 1). Temperatures from 40 to 75° F cover the range usually encountered in California streams and hatcheries, the data for the more common temperatures from 45 to 55°F being shown by intervals of 1 degree. *The marked decrease in content of dissolved oxygen with increasing temperature and altitude should be particularly observed.*

The saturation level for dissolved oxygen is less in sea water, being approximately 75% of that for the same temperature in fresh water. Intermediate salinities have oxygen saturations that lie almost on a straight-line curve between the limits of fresh water and sea water. Instructions for making dissolved oxygen determinations in fresh water are given near the end of this section.

TABLE 1
Dissolved Oxygen in Parts Per Million for Fresh Water, in Equilibrium With Air

Temperature in degrees F.	Elevation in feet										
	0	1,000	2,000	3,000	4,000	5,000	6,000	7,000	8,000	9,000	10,000
40	13.0	12.5	12.1	11.6	11.2	10.8	10.4	10.0	9.6	9.3	9.0
45	12.1	11.7	11.2	10.8	10.5	10.1	9.7	9.3	9.0	8.7	8.4
46	11.9	11.5	11.1	10.7	10.3	9.9	9.6	9.2	8.9	8.6	8.3
47	11.8	11.3	10.9	10.5	10.2	9.8	9.4	9.1	8.8	8.5	8.2
48	11.6	11.2	10.8	10.4	10.0	9.7	9.3	9.0	8.7	8.3	8.0
49	11.5	11.1	10.6	10.3	9.9	9.5	9.2	8.9	8.6	8.2	7.9
50	11.3	10.9	10.5	10.1	9.8	9.4	9.1	8.7	8.4	8.1	7.8
51	11.2	10.8	10.4	10.0	9.7	9.3	9.0	8.6	8.3	8.0	7.7
52	11.0	10.6	10.2	9.9	9.5	9.2	8.9	8.5	8.2	7.9	7.6
53	10.9	10.5	10.1	9.8	9.4	9.1	8.7	8.4	8.1	7.8	7.5
54	10.8	10.4	10.0	9.6	9.3	9.0	8.6	8.3	8.0	7.7	7.4
55	10.6	10.3	9.9	9.5	9.2	8.9	8.5	8.2	7.9	7.6	7.3
60	10.0	9.6	9.3	8.9	8.6	8.3	8.0	7.7	7.4	7.1	6.8
65	9.4	9.1	8.8	8.4	8.1	7.8	7.5	7.2	7.0	6.7	6.4
70	9.0	8.7	8.4	8.0	7.8	7.4	7.2	6.9	6.7	6.4	6.1
75	8.6	8.3	8.0	7.7	7.4	7.1	6.8	6.5	6.3	6.1	5.8

Hydrogen Ion Concentration (pH)

Because of its convenience, the pH scale [1] is quite widely used as a method of expressing hydrogen ion concentration; i.e., the acid or alkaline intensity of a water. On the pH scale, which ranges from 0 to 14, the neutral point is 7; lower values are more acidic and higher values more alkaline. The pH of natural waters will vary from about 4 to 9, the lower values being found in boggy or swampy areas, the higher ones in streams such as our western alkaline streams. Ellis et al., as the result of some 10,000 tests, state in their bulletin that normal values range from 6.7 to 8.6, and that in 90% of the areas where good freshwater fish faunas were found the pH range was 6.7 to 8.2. Values outside this range should be regarded with suspicion. Normally, the lower the pH value, the lower the mineral content. In general, waters slightly on the alkaline side support more fish than waters on the acid side. Instructions for making pH determinations will be found near the end of this section.

---

[1] pH is the negative value of the power to which 10 is raised in order to obtain the concentration of hydrogen ions in gram-molecules per liter.

Carbon Dioxide

All natural waters contain some carbon dioxide, the quantity at equilibrium being approximately 2 ppm. In well supplies, carbon dioxide may be extremely high and is usually associated with a deficiency of dissolved oxygen. Low values of 1 ppm or less usually indicate algal activity, which tends to absorb the free carbon dioxide, at the same time liberating dissolved oxygen. Low carbon dioxide is, therefore, usually associated with high oxygen values. Values above the normal of 2 ppm may be looked upon with suspicion, since they may indicate either pollution from organic waste or a deficiency of oxygen.

Alkalinity

The alkalinity of normal fresh waters results principally from calcium and magnesium bicarbonates, and is sometimes associated with potassium or sodium bicarbonate. In very alkaline waters, carbonates as well as bicarbonates may be present. Hydroxide, the most caustic form of alkali, is not found in natural waters unless caused by pollution. Both hydroxides and carbonates show a red color with phenolphthalein, whereas bicarbonates do not. For this reason, phenolphthalein is used to test for the presence of hydroxides and carbonates, and methyl orange is used for bicarbonates. While alkalinity of normal fresh waters is due principally to bicarbonates, it has become general practice in reporting alkalinity to calculate it in terms of calcium carbonate. When the alkalinity is reported as methyl orange alkalinity, only bicarbonates are indicated, but the calculation is in terms of calcium carbonate.

The quantity of bicarbonates present in most California streams varies from a minimum of about 5 ppm to about 200 ppm The higher values, being associated with larger quantities of calcium and magnesium, are considered to be more beneficial to fishlife, while streams of extremely low alkalinity may be deficient in certain mineral essentials. However, neither extremely low nor extremely high methyl orange alkalinity within the range previously stated may be considered detrimental to fish life. The determination of alkalinity is, therefore, usually for the purpose of classifying the water within a certain range or type rather than for determining whether a water can be classed as satisfactory or unsatisfactory.

Interrelationships of Carbon Dioxide, pH, and Alkalinity

A definite relationship exists between the above three water components. With a normal content of carbon dioxide, the pH increases as the alkalinity increases. Having determined the latter two, the carbon dioxide value can be computed or read from a chart without the necessity of making an analysis.

With a water that is high in carbon dioxide and, therefore, below normal pH (more acidic), aeration will cause a loss of carbon dioxide and a corresponding increase in pH (more alkaline). A simple test that may be of some value in hatcheries consists of the addition of a few drops of an indicator, such as Universal Indicator, to a small quantity of the hatchery water in a flask or bottle. The color produced is observed and then the water is agitated by thoroughly shaking the flask, to see if an appreciable color change occurs. A change in color toward the alkaline side indicates loss of carbon dioxide and the probability that the water is deficient in oxygen. A change in color toward the acid side indicates a decrease in pH, and the usual interpretation would be that the water supply is well aerated and that the carbon dioxide content is extremely low, due to photosynthetic action of aquatic plants.

The use of Universal Indicator at hatcheries for testing the water before and after aeration may be found very useful in indicating changes that occur in the characteristics of the water supply, and may at least give a clue to difficulties that develop.

If it is desired to express alkalinity in terms of bicarbonates rather than in the conventional manner as calcium carbonate, the latter figure should be multiplied by 1.22. In some reports the term "fixed carbon dioxide" is used rather than alkalinity, and this term includes all of the carbon dioxide present as normal carbonates plus half of that present as bicarbonates. When a water contains only bicarbonates, as is normally the case, the alkalinity figure in terms of parts per million calcium carbonate can be converted to the equivalent quantity of fixed carbon dioxide in terms of cc per liter by multiplying by the factor 0.18.

As previously stated, with normal carbon dioxide there is a fixed relationship between pH and alkalinity (Table 2). Lower pH values than those presented here represent either pollution or under-aeration, while higher pH values indicate good aeration and high activity of aquatic plants.

TABLE 2

Normal Alkalinities and pH Values When the Carbon Dioxide Content Is 2 p.p.m.

pH	Alkalinity in p.p.m.	pH	Alkalinity in p.p.m.
6.4	2.5	7.4	25.0
6.5	3.0	7.5	31.0
6.6	4.0	7.6	38.0
6.7	5.0	7.7	50.0
6.8	6.0	7.8	60.0
6.9	8.0	7.9	80.0
7.0	10.0	8.0	100.0
7.1	12.0	8.1	120.0
7.2	15.0	8.2	155.0
7.3	20.0	8.3	190.0

Total Dissolved Solids

The mineral content or total solids in parts per million for California streams varies between quite wide limits. The minimum content as shown in Water Supply Paper No. 237, titled "Quality of California Surface Waters", Van Winkle and Eaton (1910), is 8.1 ppm, and the maximum indicated is 1,766 ppm. The above paper is an excellent reference on characteristics of California surface waters. Present hatcheries within the State have mineral contents varying between the limits indicated above, the lowest being Moccasin Creek Hatchery with 12 ppm, and the highest, Fillmore Hatchery with 904 ppm.

Effect of Galvanized Pipes on Hatchery Water Supply

It has been demonstrated that very small quantities of zinc, copper, or lead are lethal to several varieties of fish. Zinc used in the galvanizing process will, under certain conditions, depending on the pH and temperature of the water, go into solution. Since the toxic effect of zinc in small quantities is cumulative, trout and salmon may be killed. Rainbow trout two to four weeks old have been killed in a water supply having a pH of 7.6 and containing 0.04 ppm zinc.

Small concentrations of zinc in solution, resulting from water flowing through only 15 feet of 1½ inch diameter galvanized pipe at the rate of 30 gallons per minute, have been proven responsible for heavy losses among rainbow trout fry.

In selecting pipes, flumes, and conduits for hatchery use, galvanized material must always be avoided.

Determination of Dissolved Oxygen

A number of methods for determining dissolved oxygen is available. A small kit which is clean and simple to operate is made by the Hach Chemical Co., Main Office and Factory, P. O. Box 907, Ames, Iowa 50010. Western Office, Sales-service Center, P. O. Box 477, Laguna Beach, California 92652. One of these should be available in all hatcheries. It is the Dissolved Oxygen Test Kit, Model OX-2P. For determining dissolved oxygen, Hach utilizes the Alsterberg (Azide) modification of the standard Winkler method with two important improvements. For the test solution, a time proven reagent called PAO is used. PAO does not lose strength during storage. Previously, a thiosulfate solution was used which often deteriorated quite rapidly.

The second improvement is the use of chemicals in dry powder pillows for the corrosive alkaline iodide-azide solution, manganous sulfate solution and the messy, dangerous concentrated sulfuric acid solution. Thus the kit is safe, easy to use, and the chemicals are stable; identical results are obtained as compared with the usual liquid reagent method.

Since this kit uses the drop-count titration principle, no color standards are necessary. The test is conducted in a matter of minutes with five easy steps.

Following are instructions for the use of Model OX-2P Dissolved Oxygen Test Kit.

High Range

1. Fill the dissolved oxygen sample bottle with the water to be tested by allowing the water to overflow the bottle for 2 or 3 minutes. Be certain there are no air bubbles in the bottle.

2. Add the contents of each Dissolved Oxygen 1 Powder Pillow (Manganous Sulfate) and Dissolved Oxyen 11 Powder Pillow (Alkaline Iodide-Azide). Stopper in a manner to exclude air. Shake to mix and allow the floc that is formed to settle.

3. Remove the stopper and add the contents of one Dissolved Oxygen 111 Powder Pillow (Dry Acid). Restopper and shake to mix. The floc will dissolve and a yellow color will develop if oxygen was present. This is the prepared sample.

4. Fill the sample measuring tube level full with prepared sample, and pour it into the mixing bottle.

5. Add drop wise to the dissolved oxygen bottle, swirling to mix, PAO Solution, counting each drop, until the color changes from yellow to colorless. The ppm dissolved oxygen is equal to the number of drops used.

Note: It is a bit tricky to stopper the BOD bottle without getting an air bubble trapped in the bottle. To avoid the air bubble, incline the dissolved oxygen sample bottle somewhat, and insert the stopper with a quick thrust. This will force the air bubbles out.

Low Range

Steps 1–3. Same as above.

4. Pour off contents of the dissolved oxygen bottle until the level just reaches the mark on the bottle.

5. Add dropwise to the dissolved oxygen bottle, swirling to mix, PAO Solution, counting each drop, until the color changes from yellow to colorless. Each drop of PAO added is equal to 0.2 ppm dissolved oxygen in the sample.

This Model OX-2P is available on Catalog Order Number 1469-00 for about 20 dollars and has material for making 100 tests. Additional materials are available.

There are also many types of meters available which are calibrated for direct reading of dissolved oxygen and are gradually replacing the titration methods.

pH Determinations

There are two kinds of methods for determining pH: (a) electrometric and (b) colorimetric. When highly accurate determinations are required, the electric pH meters are a must. In other cases, the colorimetric method should be adequate.

One system of determining pH colorimetrically involves the use of indicator solutions. These solutions are added to the sample and a color change occurs. The sample is then compared with a series of color standards. The sample is matched to the standard and the pH of the matched standard is the pH of the sample.

Another even simpler method uses a roll of paper which has an indicator

FIGURE 3—Pond system at Fillmore Hatchery, Fillmore, Ventura County, California. There are 10 ponds in each series. *Photograph by George Bruley, 1973.*

impregnated in it. A strip of the paper is torn from the roll, immersed in the sample, then removed and compared with paper color standards. When the test strip of paper is matched, the pH is read from the paper color standards.

This product is called pHydrion paper and is manufactured by the Micro Essential Laboratory, Brooklyn 10, New York. It is usually available from local laboratory or chemical supply houses. These papers are available in eight wide-range and 20 short-range types. In general, it is advisable to cover the range pH 5 to 9. Ranges above or below this are usually for special purposes.

FIGURE 4—Aerator midway in pond series at Fillmore Hatchery. Hatchery is operated on well water which is run through an aerator before entering the ponds, aerated between the fifth and sixth ponds, and a portion of the water returned to upper end of system to pass through another aerator before reentering the pond system to be reused. *Photograph by George Bruley, 1973.*

Recirculation and Aeration

Hatcheries should be operated at maximum capacity for the most economical production of fish. An example is the Fillmore Hatchery in Southern California where there are ten ponds in each of four series (Figure 3). The water is pumped from wells and passes through an aerator to dissipate harmful gases and increase the oxygen before entering the ponds. After the water passes through five ponds in each series, it is pumped through another aerator and then returned to the ponds to pass through five more ponds in each series (Figure 4). This additional aeration midway through the pond series increases the production of fish in the lower ponds. Some of the water from the discharge is then pumped back to the upper end of the pond system and passed through another aerator before being distributed to the pond series to be used a second time. The water temperature is 60° F. The production at this hatchery has been increased considerably by this recirculation and aeration method. The hatchery has produced an average of 15,000 pounds of catchable trout per pond in 1 year.

Reconditioning Hatchery Water

There are built-in dangers in the recirculation system described above. The oxygen can be replenished through aeration and most of the carbon dioxide can be dissipated, but very little of the ammonia can be removed.

Reconditioning the water properly infers that the oxygen will be replenished, carbon dioxide and nitrogen will be dissipated, and ammonia will be removed so the water can be recirculated many times by adding only 2 to 10% new water.

Mad River Hatchery in Humboldt County north of Eureka (Figure 5) operates on reconditioned water that is recirculated. The total water requirement is 30 cfs of which only 3 cfs, or 10%, is supplemental well water which is added to the system. When water is wasted for reasons mentioned below, it is necessary to increase the supplemental water.

FIGURE 5—Mad River Hatchery, Arcata, Humboldt County, California. Hatchery operates on 30 cfs of water, 90% of which is reconditioned. Only 3 cfs of supplemental water is added to the system. *Photograph by George Bruley, 1973.*

Ammonia is the principle catabolic product which causes difficulty in water reuse systems. Ammonia can be removed by nitrifying bacteria. In nitrification, ammonia is first converted by *Nitrosococcus* and *Nitrosomonas* to nitrous acid, then by *Nitrobacter* to nitric acid. The acid combines with an available base to form nitrites and nitrates. The ultimate product is nitrate which is harmless in the recirculating water.

Bacterial beds for the culture of nitrifying bacteria are developed in filters for reconditioning of hatchery water by covering a 4 foot deep layer of sharp rock with a 1 foot layer of crushed fresh ocean oyster shells. The oyster shells on top of the rock serve as the filter material to remove solids from the recirculated water and as a source of nutrients for the nitrifying bacteria.

The nitrifying bacteria are chemosynthetic autotrophs in that they are capable of taking the simple inorganic compounds (ammonia, oxygen, carbon dioxide, carbonate, and bicarbonate) and synthesizing them into carbohydrate, fat, and protein, using the oxidation of ammonia as their energy source. They also require other chemicals including a base for the formation of nitrites and nitrates, and phosphates, magnesium, iron, and other trace minerals. It would appear that nitrifying bacteria are practically self-sufficient in that growth can be either retarded or completely inhibited by the absence of the essential chemicals. The oyster shell supplies principally calcium bicarbonate which provides the base for production of calcium nitrate. Without the base available, nitrous and nitric acids

FIGURE 6—There are eight filter beds at Mad River Hatchery. Bubbles can be seen on the surface of the two ponds in the foreground as air surges up through the filter as part of the cleaning process. *Photograph by George Bruley, 1973.*

FIGURE 7—Reconditioning water at Mad River Hatchery. A, air surging up through water as filters are cleaned; B, a maze of pipelines in an underground corridor delivering water to and collecting water from eight filter beds and a pipe supplying air to each filter; C, air compressor to provide air to clean filters; D, a double outlet at lower end of pond series so water can be directed to filter beds for reconditioning or be wasted to settling ponds when ponds are being cleaned, or chemicals are used in ponds. *Photographs by George Bruley, 1973.*

are formed which would reduce the pH of the water. The calcium carbonate provided by the oyster shell tends to stabilize the pH at a satisfactory level.

At the Mad River Hatchery there are eight filter beds 20 feet by 75 feet (Figure 6). The size of the filter is calculated on the basis of 1 gallon per minute of recirculating water per square foot of filter surface. The water to be recirculated from the rearing ponds enters the filter ponds and is held at the normal operating level. The water passes through the filter bed, is collected in the filtered water manifold at the bottom of the filter and then pumped under pressure through aspirators into the aeration tower. The reclaimed water is then returned to the rearing ponds by gravity flow.

The filters must be cleaned at least twice each week. This is done by closing off the inflow of water and draining the filter pond down to about 12 inches in depth. The water is drained through the waste pipe. An air blower is activated for 1 to 2 hours. The air surges up through the rock and oyster shell at the rate of 1.33 cfm per square foot of filter area, agitating the oyster shell which cleans it without disturbing the rock layer (Figure 7). During the last 20 minutes of this operation, the raw water from the rearing ponds is turned in to flush the debris into the waste pipe. The filter pond is then filled and ready for operation. If algae becomes a problem, it is sometimes necessary to rake the oyster shell during the cleaning process.

Water should be wasted from the system when rearing ponds or filter beds are being cleaned or when any chemicals are used to treat fish in the ponds (Figure 7D). Chemicals might kill the culture of nitrifying bacteria. Additional water must be added to the system at that time.

Waste Water Discharge Requirements

Water quality standards for waste water discharge from hatcheries have been strengthened by the U. S. Environmental Protection Agency and the State Water Resources Control Board. The Federal authority stems from the 1899 Federal Refuse Act which prohibits discharges into navigable waters of the United States and their tributaries without a permit issued by the Army Corps of Engineers. At this writing it appears that this authority will pass to the state level which was made possible by the Federal Water Pollution Control Act amendments of 1972.

Indications are that dischargers will be required to install at least secondary treatment by 1977 or the best practical control technologies available. Standards will probably be set that no discharge can degrade the quality of the receiving water. In some cases the requirements may be met by passing the discharge water through large settling ponds. By 1983, it is probable that fish hatcheries will have to operate on a closed system of reconditioning the water and reusing it as in the system previously described at Mad River Hatchery. Strict standards would be set for the water that is wasted from such a hatchery before it could pass into a public water.

STRUCTURE OF THE TROUT EGG

Which came first, the egg or the fish? Rather than debate this subject here, we will turn to the hatchery which does not produce its own eggs. Here the egg comes first.

While much has been learned and written regarding the taking and care of fish eggs, a great deal remains to be learned. In comparing the work at many hatcheries and egg-taking stations, it becomes apparent that there is a great lack of uniformity insofar as egg-taking operations are concerned. Every fish culturist knows that eggs are delicate and require constant and gentle care. Certain temperature limitations must be maintained, smothering must be guarded against, rough handling or shock must be avoided, and fungus, *Saprolegnia,* must be kept under control. It is no secret that eggs which turn white are dead, but what makes them turn white? In order to better understand this subject, we will review a few points pertaining to the general structure of the trout egg.

The shell of the trout egg is porous. The minute pores can be seen with a low power compound microscope. The shell is also elastic and varies in strength and thickness among eggs from different varieties of fish and even from different females of the same variety. One cannot speak of water circulating through the shell, but there is no doubt whatever that it can diffuse through slowly. In general, the highly transparent shell, which allows a fairly good view of the developing fish, is thin. The thick shells are whitish and opaque. It is not known which type of shell is more desirable from the hatchery point of view, or whether the thickness can be influenced by diet. It appears to be an hereditary matter.

The shell contains one larger opening, the micropyle, through which the spermatozoon enters to fertilize the egg (Figure 8A). This opening was first seen in a trout egg by a German scientist named Bruch, in 1896. It is quite simple, consisting of a little hole at the base of a crater-like depression in the shell. The shell is not thinner around the hole than elsewhere, merely depressed. In response perhaps to some chemical attraction from within the egg, the spermatozoon enters the micropyle and penetrates the yolk membrane to unite with the nucleus of the egg. It is a curious fact that although a number of spermatozoa may enter the micropyle, only one of them succeeds in uniting with the egg nucleus. Physiologists think this is due to a chemical phenomenon, and that the entrance of the spermatozoon into the micropyle is not the result of any chemical attraction but merely an accident that depends upon the presence of a very large number of spermatozoa. It might be mentioned that the mechanism of fertilization is a problem over which battles are still being fought by scientists.

The yolk membrane, mentioned above, is a protoplasmic layer which surrounds the yolk and holds it together. This membrane is very thin, comparable to that we see holding the yolk of a hen's egg together when we break one into a frying pan. It is not porous like the shell. The yolk membrane is of great importance to the fish culturist, for it is the rupture of this delicate layer which causes an egg to turn white and which is responsible for most of the losses which occur among eggs in hatcheries.

In a trout egg, as we see it in a hatching trough, there is a space between the yolk membrane and the shell and this space is filled with a fluid which embryologists call the perivitelline fluid. When an egg is freshly stripped from the female, there is no water in this space; in fact, the space itself hardly exists. The outer shell is quite limp, and the whole egg is rather collapsed, or flaccid. As the egg absorbs water it becomes swollen, and firm or turgid (Figure 8B). Eventually it absorbs 20% of its initial volume in water. This swelling is part of the "hardening" process familiar to all fish culturists. It was formerly thought that this absorption of water created a current through the micropyle which carried the spermatozoon into the egg. But we know now that this could hardly be a factor in the fertilization of the egg, for fertilization is effected within a few seconds after

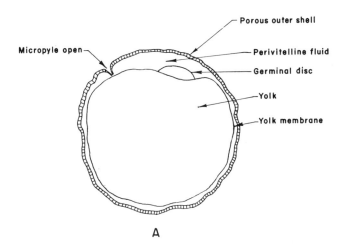

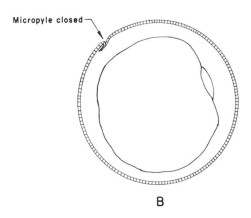

FIGURE 8—A, newly taken trout egg, outer shell is not firm, egg feels soft and slightly adhesive; B, water-hardened trout egg, outer shell is drawn tight, micropyle is now closed, egg is firm and slick.

the spermatozoon and the egg meet, whereas water is absorbed over a period of 20 minutes or more, so that there could hardly be much of a current through the micropyle. Furthermore, high fertilization can be obtained by the dry method, in which no water is added until after the eggs and sperm have been mixed.

The reason why the egg absorbs water into the space between the shell and yolk membrane until it becomes turgid is explained as follows. When the egg is first placed in water a slight shrinkage of the yolk mass occurs as a result of the extrusion of a colloidal substance into the perivitelline space. This colloid has the property of taking up water, and upon so doing its volume is considerably augmented. The pores of the shell are so fine that although water and the class of soluble substances known as electrolytes can pass through, the colloids can not. Therefore, the colloids remain within the perivitelline space and continue to absorb water until their power to do so is counteracted by the resulting pressure of the shell. The stage of equilibrium is reached within about 20 minutes, and thereafter no further changes take place in the volume of the egg.

The outer shell itself toughens during the hardening period. This change is brought about through absorption of water into the material of the shell.

On the surface of the yolk, at the time the egg is ejected, is a quantity of protoplasm within which lies the egg nucleus. As the egg hardens, this protoplasm migrates toward one spot, where it gathers just beneath the yolk membrane into a small raised mass called the "germinal disc". This occurs whether the egg is fertilized or not, although the mass is never quite as high and compact in an unfertilized egg as in a fertilized one. It requires some hours after the hardening process is complete for the disc to reach its maximum height.

The yolk is a viscous, yellowish fluid which contains quite a number of oil droplets. These droplets congregate in the upper portion of the yolk and probably help maintain the germinal disc at the top of the egg. The whole yolk mass can rotate freely within the shell. If an egg is turned until the germinal disc is at the side, the yolk soon rotates and carries the disc back to the top of the egg. Therefore, the micropyle, although it must be located near the germinal disc in a newly-ejected egg, bears no relation whatever to the germinal disc once the egg is water hardened.

We are now ready to return to our original question of why a dead egg turns white. The yolk contains a good deal of protein material called globulin. The globulin is in solution in the yolk and is held in this state by the presence of salts. It cannot remain in solution in water which contains no salts or other electrolytes, so that removal of salts from the yolk causes it to precipitate. In a normal egg the yolk membranes prevent the salts from diffusing into the perivitelline space and from there through the porous shell. However, if the yolk membrane disintegrates or breaks and allows the salts to leak out, the globulin precipitates and turns white. Whiteness is caused by precipitated globulin.

If a white egg is placed in a weak solution of salt (about 1%), it will slowly regain its normal color. The globulin, because of the absorption of salt from the water, will again go into solution. If such eggs (kept in salt water) are examined under a binocular microscope, the cause of the whitening can usually be seen in the form of a break or hole in the yolk membrane. If the egg is replaced into fresh water, again the absorbed salt will leach out, the globulin will precipitate, and the egg will turn white.

An examination of the cleared egg will also disclose whether the egg was sterile or whether it had undergone some development prior to the rupture. Both kinds will be found among white eggs.

During the development of the embryo in the fertilized egg, the yolk membrane soon becomes covered by a much thicker wall of cells, so that after a while it is no longer in any danger of being broken. Until this occurs the egg is still tender. Naturally, a sterile egg never passes out of the tender stage. The familiar hatchery practice of agitating the eggs after they are eyed, called shocking or addling, ruptures the yolk membranes of the ever-tender sterile eggs. The result is a precipitation of the globulin and a whitening of the egg.

Little is known regarding the conditions which cause yolk membranes to rupture. Mechanical shock can rupture the membranes very easily. Every fish culturist is also familiar with the fact that when a white egg becomes fungused, and is not removed from the tray, the fungus gradually envelopes the neighboring live eggs and after a while they also turn white. The fungus probably makes a chemical attack upon the membrane. No doubt chemical injury to the yolk membrane could also occur in certain polluted waters.

SPAWNING (EGG TAKING)

The act of obtaining eggs from female fish and sperm from male fish is referred to as spawning, egg taking, or stripping. The words egg and ova are synonymous. Even though the merits of artificial propagation are not as great as first claimed in the early annals of fish culture, the efficiency of artificial fertilization can hardly be disputed and fertilization of nearly 100% has been attained in trout eggs by artificial methods.

FIGURE 9—A modern brood pond and spawning house at Mt. Shasta Hatchery, Siskiyou County, California. *Photograph by George Bruley, 1973.*

Spawn-Taking Facilities

It may safely be said that the less the fish are handled the better, and streamlining spawn-taking operations to reduce handling is certainly a step in the right direction. The modern spawning house is arranged so the fish are not carried into the spawning house, but instead are herded into the spawning enclosure without being taken from the water.

FIGURE 10—Openings are provided for each brood pond into the center flume so fish can be herded between ponds or into the spawning house and returned to any pond without removing them from the water. *Photograph by George Bruley, 1973.*

Three or four concrete ponds 10 feet wide and 100 feet long in series are used for trout brood fish at state hatcheries. These ponds are built in pairs with a 3 foot concrete flume between them (Figure 9). There are openings into each pond from the center flume so fish can be herded into the spawning house from any pond and returned to any pond without removing them from the water (Figure 10). Doors on the spawning house that can be lowered into the water make it possible to move the fish inside and keep the spawning house reasonably warm which greatly facilitates the spawning operation (Figure 11). One can understand that unless a person's hands are kept reasonably warm, it is difficult to feel the amount of pressure being applied to the fish in expressing the eggs. The egg taking operation is speeded up under comfortable conditions, resulting in time saved, better fertilization and less injury to the fish. Perforated screens can be placed in various locations to provide compartments for use in sorting and spawning the fish. The fish are dipped up a few at a time and anesthetized by placing them in an anesthetic solution for a short time (Figure 12). The anesthetic commonly used is MS222. Anesthetized fish are much easier to handle for sorting and there is less danger of breaking eggs or injury to the fish in spawning when the fish do not fight and struggle. The fish recover and regain their equilibrium very rapidly when returned to fresh water.

Salmon and steelhead leave their parent stream, ascend a fish ladder (Figure 13), and pass through a V-trap into a concrete gathering tank. A mechanical crowder (Figure 14) with a vertical bar rack which can be raised and lowered hydraulically runs on the concrete walls of the gathering tank. This crowder directs the fish into the spawning house. The most recent model is operated electrically by pushing a button in the spawning house to move the fish in as required. The fish are pushed into a metal basket which is lowered into an anesthetic tank (Figure 15). The basket is raised after the fish are anesthetized and the fish are worked onto the sorting table. When the fish are sorted for sex and degree of ripeness, the green fish are sent to various holding tanks through metal flumes and pipes (Figure 16) or herded with push racks in flumes full of

FIGURE 11—Brood fish being moved into the spawning house for sorting and spawning. Note the overhead door that can be lowered into the water to keep the spawning house warm during winter months (also shown in Figure 9). Pens can be made with crowding racks for use during the sorting and spawning operations. *Photograph by George Bruley, 1973.*

FIGURE 12—Fish are dipped a few at a time into a container with an anesthetic before sorting and spawning. *Photograph by George Bruley, 1973.*

water. Circular tanks with gates to a central flume are used in some facilities (Figure 17). Rectangular ponds are used in other facilities (Figure 18). The rectangular ponds have the advantage of greater ease in pushing the fish with an electrically operated rack into the gathering tank where the mechanical crowder can herd them into the spawning house for further sorting and/or spawning. There is more manpower involved in guiding the fish out of the circular ponds.

The water circulation in the round ponds is as described in the section on circular ponds. The water circulation in the rectangular ponds also operates in a similar way. Two rectangular ponds have a common center wall which does not extend to the end of the ponds so the water can circulate in the two ponds around the center wall. When fish are to be removed from the ponds, mechanically operated racks are run the length of the two ponds simultaneously.

The mechanical devices have reduced the labor required in sorting and spawning salmon and steelhead as well as causing less injury to the fish. Similar mechanical devices are used in handling trout brood fish and are a definite advantage.

Spawning Season

The spawning season of trout varies to some extent with the locality and the temperature of the water. By nature, brook and brown trout are fall spawners, while rainbows spawn in the spring (a complete list of the common and scientific names of the salmons and trouts of California is contained in the section on classification of fishes). In some California streams with runs of anadromous fishes, such as the Trinity River, there is hardly a month in the year when some strain of fish is not spawning. As an example, spring-spawning king salmon arrive in the upper river section at Lewiston the latter part of June. They are then followed by the summer and fall runs, which extend the salmon spawning season to mid-November. The king salmon runs overlap the silver salmon runs, which extend from November through January. The silvers in turn overlap the steelhead, which spawn during the period January through June. Actually, the steelhead of the Trinity River may be divided into those of the spring run (fish in general entering and migrating upstream on dropping stream levels while quite green, and spawning in the following season), and those of the fall run (fish in general entering on rising stream levels with sexual products in various stages of development, but spawning within the same season).

There are several ways in which the spawning season can be advanced. These include selective breeding, use of artificially controlled light, and injection of pituitary hormones. Selective breeding has been highly effective, and through this process the fall-spawning rainbow commonly referred to as the Hot Creek strain

FIGURE 13—A typical fish ladder for salmon and steelhead at Iron Gate Hatchery, Hornbrook, Siskiyou County, California. *Photograph by George Bruley, 1973.*

FIGURE 14—A mechancial crowder that runs on the walls of the gathering tank pushes the salmon and steelhead into the spawning house. Note the V-trap in the center of the picture where the fish enter the gathering tank from the ladder. The rack on the crowder is raised and lowered mechanically. *Photograph by George Bruley, 1973.*

has been developed. Even though selective breeding is a very slow process, the results not being evident until several generations of offspring have been dealt with, the results are permanent and once the strain of fish has been developed it can be perpetuated by proper selection. It has been clearly demonstrated that the spawning time of trout can be advanced by the injection of pituitary hormones. It offers interesting possibilities for crossbreeding varieties which spawn at different times of the year.

Effect of Temperature on Brood Stock and Eggs

Probably no other single thing regulates the development of eggs and growth of fish as much as temperature. It has been shown that rainbow trout reared in water held fairly constant at 60° F grow in length at the rate of about 1 inch per month. At 45° F they grow at less than ¼ inch per month.

FIGURE 15—**A**, fish being pushed from the gathering tank into the anesthetic tank to be anesthetized; **B**, the basket is raised after the fish are anesthetized, ready to be put on the sorting table; **C**, sorting table on the right center of picture. *Photographs by George Bruley, 1973.*

FIGURE 16—Unripe fish can be sent to the holding tanks through metal flumes. **A and B**, metal flume from spawning house to the head of rectangular tanks; **C**, trap door where fish enter the pond and electrically operated crowding racks to push fish through rectangular ponds and into the gathering tank. *Photographs by George Bruley, 1973.*

It is quite generally agreed that yearling and adult rainbow can withstand temperatures up to 78° F for short periods of time without harmful effect. It has also been shown that in order to produce eggs of good quality, rainbow spawners must be held at water temperatures not exceeding 56° F, and preferably not above 54° F, for a period of at least 6 months before spawning.

In order to get rainbow brood stock to grow rapidly and spawn when 2 years old, it is common practice to rear them the first 16 months of their life at a location where water temperatures are fairly constant at 60° F, and then to transfer them to a location with a water temperature below 54° F to mature.

Just as water temperatures which are too warm (higher than 56° F) adversely affect egg development in rainbow and king salmon spawners, so do water temperatures which are too cold (42° F or lower) affect the development and incubation of trout and salmon eggs. In one experiment in which mature adult king

FIGURE 17—Salmon and steelhead are held in circular tanks at Iron Gate Hatchery. There is a flume on each side of the spawning house, each connecting with three tanks. Push racks are used to move the fish to these tanks. Note the fish ladder in the background and the gathering tank between the upper end of the ladder and the spawning house. *Photograph by George Bruley, 1973.*

A

B

FIGURE 18—**A**, rectangular ponds are used for holding salmon and steelhead at Mad River Hatchery. Note metal flume at right of picture to return fish to ponds from the spawning house; **B**, the lower end of the rectangular ponds as they connect to the gathering tank, every second wall dividing the ponds does not extend the full length of the pond so two ponds can operate as a circular type pond. The fish ladder from the river can be seen in the background connecting to the gathering tank with a V-trap and the crowder. Note reel for the electrical cable that provides current to operate the crowder. *Photographs by George Bruley, 1973.*

salmon females, nearly ready to spawn, were placed in water ranging from 34 to 38° F, none of the females ripened and all died before spawning.

In an attempt to incubate king salmon eggs at a constant 35° F temperature, mortality was practically 100%. Eggs held at water temperatures of 42.5° F or higher developed with only normal loss. Salmon eggs which have been held in water slightly above 42° F for a period of 6 days or longer could then tolerate colder temperatures without excessive mortalities. It is safe to say that the eggs of rainbow trout and king salmon will not develop normally in the fish if constant water temperatures above 56° F are encountered. It also follows that both rainbow trout and king salmon eggs cannot be incubated in water below 42° F without excessive loss.

Care of Brood Stock

As in all animals, the time of sexual maturity is one of the most critical during the life of a fish, so brood fish should be handled with the utmost care.

In order to obtain eggs of the highest quality, brood fish must receive an adequate diet. As more of the secrets of trout nutrition are learned, better diets and improved egg quality may be expected. Fish culturists are cautioned against using any diet containing regular cottonseed meal for brood fish. If fed in more than slight amounts, it will adversely affect egg production in fish, just as it affects the reproductive system in other animals. Apparently the effect of degossypolized cottonseed meal on the quality of trout eggs is unknown.

A female trout, heavily laden with eggs, cannot withstand the rough handling sometimes associated with poor hatchery practices. Great care should be taken during the sorting and spawning operation to dip up only two or three fish at a time. *Never make a pass through a pen of nearly mature females and fill the bag of the dip net with fish.* This can result in broken eggs, poor fertilization, and possibly permanent injury to the fish's reproductive system.

Selective Breeding

Improvement of animal and plant life of various types through selective breeding is an old art, which has produced nearly all of our domestic animals and cultivated plants. In many cases the evolution of our present strains and varieties is almost lost in antiquity. In recent years, the new science of genetics, applied to plant and animal breeding, has greatly accelerated the rate at which improvement can be made. Improved strains of hybrid corn, for example, have been developed rapidly.

In countries where fish farming is important, special strains have been developed for various purposes, including rapid growth, disease resistance, greater beauty, etc. There is no question that nature exercises selection in the survival of animals in various wild environments.

To meet the demands of an artificial operation, such as our present hatchery program, there are many things which must be considered to make the program a success. Production of eggs at the right time of year, so the resulting fish will reach correct size for stocking when they are needed, is highly important. The development of strains which mature at 2 years of age, instead of 3, saves the cost of holding brood fish an extra year. Selection for large eggs gives the resulting fry a better start in life and higher survival.

The selection of females producing a large number of eggs is important, since it reduces the number of brood fish required. It is always important to select eggs of high quality. It is important to select the largest fish from a 2 year old strain, since this will insure a rapidly growing fish which will increase the economy of the production program. it is essential to select fish with good conformation and coloration in both the male and the female; otherwise, deformities and poorly-shaped fish will be produced. Disease resistance also can be developed through selective breeding.

In California, selective breeding has been carried on since 1938. It is of interest to note that the fall spawning rainbow were developed over a period of many years, beginning about 1883, when eggs were taken from wild spring spawning rainbow from the McCloud River, California, by the U. S. Fish Commission and shipped to its hatchery in Neosho, Missouri. After many years of selection at Neosho, some of the fish were shipped to its hatchery at Springville, Utah, where further selection for early spawning was made. As a result of continuous selection, these normally spring spawning rainbow became fall spawners. A shipment of eggs of these fall spawning fish obtained for Hot Creek Hatchery in 1933 formed the nucleus of our present stock.

Through selective breeding of the Hot Creek strain of rainbow trout, 2 year old spawners have been developed in three generations. The number of fish spawning at 2 years of age has been increased from 53% to 98%. The egg production from 2 year old females has been increased fourfold in six generations. Weight of the selected fish at one year of age has increased from 5 ounces each to 10.2 ounces in five generations.

Selective breeding has developed strains of rainbow trout which spawn in all months of the year except May or June. Through continued selective breeding, it will be possible in the future to produce eggs in all months of the year in the amounts required for a year-round hatchery program.

Most of the selective breeding in California has been with rainbow trout at four hatcheries holding broodstock; although, other strains of brood trout held at hatcheries have also been bred selectively.

The production of hybrid trout, through crossbreeding of different species, has great promise but should be evaluated very carefully to be sure a desirable strain is developed.

The California Department of Fish and Game appointed a Trout Broodstock Committee in 1953 to guide the selective breeding program. Although the gains have been significant in the past, it became evident that future improvements were dependent on the development of a systematic selection procedure and an efficient system of mating. The mating system required was one which maximized the effects of selection but, at the same time, minimized the rate of inbreeding.

A detailed report, "Rainbow Trout Broodstock Selection Program with Computerized Scoring", was prepared by Graham A. E. Gall, Geneticist, Department of Animal Science, University of California, Davis, in conjunction with the Trout Broodstock Committee in 1970 (Gall, 1972). California fish culturists should follow the directions in this full report or any revision in conducting selective breeding programs. The following is only a summary of this program.

The objective of the selective breeding program is to supply quality brood fish and at the same time continue to improve the economics of the production of fingerlings, subcatchable and catchable size fish through genetic means. In the program's present form, each broodstock population will be selected for attributes considered necessary for improvement of the net merit of the stock. The attributes and their relative values are:

Attribute	*Relative Value*
Size of eggs:	
Female parent	0.2
Full sisters (random sample)	0.2
Number of eggs:	
Female parent	0.45
Full sisters (random sample)	0.45
Percent of egg mortality	0.4
Size of fingerlings	0.9
Percent fingerling mortality	1.0

FIGURE 19—Eggs from select brood females in individual hatching trays in troughs for observation and final selection. *Photograph by Jerry Eskew, 1973.*

THE RESOURCES AGENCY OF CALIFORNIA
DEPARTMENT OF FISH AND GAME
BROODSTOCK SPAWNING RECORD

Variety_____

Group _____   Hatchery_____

Age at spawning _____   Date and year _____

Lot No.	Date of spawning	Family mark		Weight		Volume of eggs (ml)	Green egg size	Number of eggs				Fingerling			Remarks
		Female	Male	Female (oz)	Male (oz)			Retained	Dead	Hatch date	Date	Size no./oz.	(oz) Total weight		

FIGURE 20—Broodstock spawning record.

THE RESOURCES AGENCY OF CALIFORNIA
DEPARTMENT OF FISH AND GAME
RANDOM SAMPLE DATA SHEET

Variety_____   Hatchery_____

Age at spawning _____   Date and year _____

Family mark	Date _____		FEMALE #2		Date _____		FEMALE #4		Date _____		FEMALE #6		Date _____		FEMALE #8		Date _____		FEMALE #10
	FEMALE #1				FEMALE #3				FEMALE #5				FEMALE #7				FEMALE #9		
	Vol.*	Size*	Vol.*	Size*	Vol.*	Size*	Vol.*	Size*	Vol.*	Size*	Vol.*	Size*	Vol.*	Size*	Vol.*	Size*	Vol.*	Size*	Vol.*	Size*

* Vol.—Ounces of eggs per female (or milliliters)
  Size—Number of eggs per ounce

It is only necessary to record data for those females not recorded on Form FG 732.

FIGURE 21—Random sample data sheet.

The attributes, percent fingerling mortality, size of fingerlings, and number of eggs spawned are given the highest relative values. The expense of raising fingerlings is high, thus losses should be kept as low as possible. The cost of fingerlings is greatly affected by growth rate, so rapid fingerling growth is desirable. The expense of maintaining brookstock can best be minimized by obtaining the largest possible number of eggs per female. The size of eggs and percent of egg mortality are given lesser values. It is felt that large eggs are important primarily in maintaining egg quality and that further improvement in lowering the percent of egg mortality would be difficult because of environmental effects.

The net merit of an individual family of fish is determined by a selection index. The index is derived by scoring each family for each attribute on the basis of its performance and weighing the score by the relative economic values assigned to that attribute. A score for the spawning performance of full sisters of the dam of the family is included in the index.

The generation interval is held at 2 years. Consequently, two sublines will develop for each broodstock. These will be maintained and no effort will be made to cross mate between the two sublines.

Each year the broodstock hatcheries spawn 60, 2 year old females with 60, 2 year old males. The 60 lots of eggs are kept separate in incubator trays or separate trays or baskets in a trough (Figure 19). A few extra lots should be taken in case some lots fail to survive. Failure to survive means if the loss to the fingerling stage exceeds 85%.

A selection is always made of at least 10 families. All fish in the 10 families have distinguishing marks. When the selection is made, no more than six females from one family should be used and mated with males of different families so there will be no sister-brother matings. To reduce the rate of inbreeding, no more than two females from a family should be mated with full brothers from another family.

A reliable source of information concerning female characteristics can be obtained by studying the performance of a random sample of females from each family. A volume measurement and a size count of water hardened green eggs from a random sample of 10 females from each family should be made. These counts can preferably be made during several weeks of the spawning season. This information with that provided from the 60 lots of eggs from 10 families will provide information to compare performance changes in the broodstock in several generations.

If it is desirable to make the spawning time earlier in the year, the 60 lots of select eggs should be taken in one or two consecutive spawnings early in the season before the peak of spawning is reached. By the same token, if it is desired to delay the spawning season, the 60 lots of eggs should be taken after the peak of spawning. If it is desired to widen the peak, a portion of the 60 lots should be taken on each side of the peak.

When fish are sorted for broodstock selection, the fish to be discarded are those which are not vigorous, healthy, and free of physical defects; or exceptionally small fish, fish having an exceptionally small number of eggs, or eggs smaller than 650 per ounce.

After the 60 lots of eggs hatch, each lot is placed in a separate compartment in a trough. When the first lot of fingerlings reaches 25 fish per ounce, all lots of fish are counted and size of fish determined.

Form FG 732, Broodstock Spawning Record (Figure 20), and Form FG 732B, Random Sample Data Sheet (Figure 21), are used to record data for computer calculation of the scores.

It is imperative that data on each lot be complete. No lots are discarded for any reason unless the mortality exceeds 85%. One purpose of the scoring system is to allow comparisons from year to year. This comparison cannot be done on the 10 lots finally selected as future broodstock. It must be done on the basis of the average of all lots which survive, preferably all 60. A comparison based on only the 10 lots selected in any 1 year would not consider the range in level of performance which could and will vary from year to year. Basically, the 60 lots spawned in any 1 year represent a sample of the broodstock being used in that year and thus yield unbiased and consistent estimates of the level of performance in that year.

A scoring system has been developed as a method of making the final selection of fingerlings for future broodstock. The system involves the assignment of numerical scores for degree of excellence for each of the attributes considered. Each lot of fingerlings is scored for each attribute and the sum of all scores is used to form an index for each lot. The lot with the highest degree of overall excellence will have the largest index value. Final selection is based on this index. A computer program has been prepared to facilitate calculation of the scores.

Ten lots are selected as future brood fish on the basis of their index values. An effort must be made to control the rate of inbreeding. It is possible that of the top 10 families (lots) selected, six could be from females which are full sisters. Therefore, a limit of three lots from any one full-sister female group would be included in the ten lots selected. If this happened, lots which rank below tenth on the index score would be selected.

All fingerlings of the ten lots finally selected are marked with a distinctive family mark. They are then all put into one group and raised as brood fish for spawning when they reach 2 years of age.

FIGURE 22—Sorting brood fish for ripeness. Sorting trough is waist high. Fish are placed in trough a few at a time after anesthetizing. Females with eggs are dropped gently into holding pens, or slid across a chute as shown, to more distant pens. *Photograph by George Bruley, 1973.*

Sorting Frequency

The interval at which brood females should be sorted (Figure 22) during the spawning season varies among hatcheries, and depends to a large extent on water temperature and season. Normally, it is not necessary to sort the females as often early in the season as during the peak of the operation. To produce eggs of the best quality, it is necessary to watch the brood stock closely. The correct degree of ripeness must be attained in the females. Taking eggs before they are fully mature (ripe) is as bad as not sorting frequently enough, which may allow some of the females to overripen. Under the unnatural conditions associated with domestication, fish rarely deposit their eggs of their own accord. If they are not sorted often enough, overripe eggs are sure to be found. Such eggs are hard and glassy, probably as a result of contact with a serous ovarian exudation present in the fish, and are sometimes referred to as "mooneyes". Besides being infertile, they may injure or even kill the fish.

Research has shown that the ripening of trout eggs can be represented graphically as a curve with a sharp apex. The peak of this curve represents the time of optimum fertility of a particular lot of eggs, which must be stripped at that time. If taken prior to this date, lower fertility results, due to the eggs not being completely ripe. If taken later, on the down side of the curve, overripe eggs are encountered. Correct timing, through proper and frequent sorting, is one of the greatest secrets of successful egg taking.

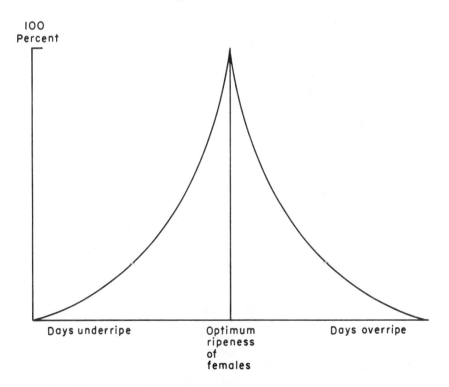

FIGURE 23—Effect of female ripeness on fertilization of hatchery trout eggs.

Size of Eggs

In general, the size of the egg depends upon the size and age of the parent fish, the larger specimens producing more and larger eggs. Egg size also varies among different strains of domestic broodstock, and among wild fish in different waters. It is reasonable to assume that competition among fry gives the larger fry a better chance for survival and faster growth. Hence, in selecting broodstock there is some advantage in selecting for larger eggs. Size, however, can be attained only at the expense of number. There is, therefore, some point at which, on the average, the forces favoring size are balanced by those favoring number. As a result of selective breeding, the eggs from domesticated stock are usually larger than those from wild fish. Mt. Whitney Hatchery spring-spawning rainbow broodstock average 427 eggs per ounce and a female produces 3.9 ounces or about 1,553 eggs when 2 years old. The same fish at 3 years of age produces about 9 ounces of eggs, averaging 254 per ounce or 2,210 eggs. This indicates that the size of the eggs increases by 40% between the second and third year of the female's life and the number of eggs produced increases by 42% (Table 3). This further points up the necessity of proper diet and handling of broodstock to insure their productive capacity over a period of several years.

TABLE 3
Record of Trout and Salmon Eggs Taken at California Stations

Variety	Location	Age	Number females spawned	Number eggs per ounce			Number ounces per fish	Number eggs per fish
				Smallest	Largest	Average		
Rainbow, spring spawn	Mt. Whitney Hatchery	Domestic, 2 years	2,570	540	350	427	3.9	1,553
Rainbow, spring spawn	Mt. Whitney Hatchery	Domestic, 3 years	2,805	311	207	254	9.0	2,210
Rainbow, fall spawn	Hot Creek Hatchery	Select, 2 years	30	490	420	447	5.78	2,600
Brown	Mt. Whitney Hatchery	Domestic, 2 to 4 years	618	290	225	249	5.65	1,751
Golden	Cottonwood Lakes	Wild	1,360	410	370	393	1.0	382
Steelhead	Snow Mountain Station, Eel River	Wild	461	240	200	221	18.75	4,304
King Salmon	Fall Creek Station, Klamath River	Wild	489	80	73	76	38.53	2,902

Anatomy of the Female Trout

Before going deeply into artificial spawning, let us look at the anatomy of the female trout (Figure 24).

Some knowledge of the anatomy of the reproductive system will guide the spawntaker and provide a scientific basis for the job. Mature male and female trout have glands called gonads on each side of and above the digestive organs. The male has testes which produce individual spermatozoa, and the female has ovaries which produce eggs. Trout lend themselves better than most fish to artificial reproduction because the eggs are loosely held in a thin membrane and can easily be manipulated toward the vent. However, eggs may be forced into the body cavity by awkward or careless practices. Such eggs can never be extruded and are eventually absorbed by the tissues.

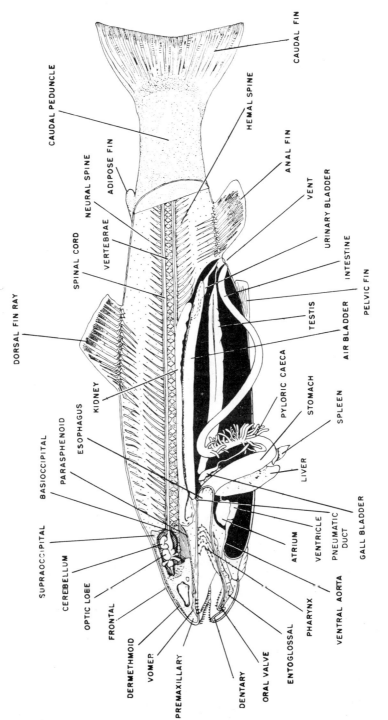

FIGURE 24—Anatomy of a trout.

Eggs, when taken, are only slightly adhesive and on extrusion and absorption of water become firm and slick.

If a fully mature (ripe) female trout or salmon is held by the tail, head down, the mass of eggs sags visibly toward the head and free eggs may settle into the forward end of the abdominal cavity outside the ova-containing membrane. The danger of this happening is even greater after the stripping process has begun and the tense condition of the supporting abdominal wall is relaxed. *Avoid holding a ripe female by the tail, head down.*

Forcible pressure in stripping may rupture the membranes and injure the ovaries, and result in a lowered egg survival. Inasmuch as trout, in the state of nature, do not emit all of their eggs at one time, no forcible attempt should be made to expel more eggs than those which flow easily under gentle pressure. It may take several manipulations to obtain the eggs. Since the eggs in the posterior part of the ovary, i.e., those nearest the vent, are the first to ripen, it is not necessary to apply pressure the whole length of the abdomen. Ripe female trout should be held tail down with the head high. This permits the eggs to flow or roll along the oviduct toward the vent.

Adhesiveness

Newly taken trout eggs are slightly adhesive and this condition continues until they are water-hardened. It is due to the water filtering through the outer shell and not to any sticky substance on the egg. It may be compared to a depressed tennis ball which has been punctured. If one holds his thumb over the puncture during the time air is attempting to fill the ball, the latter will have a tendency to cling to one's thumb. It is understandable that an egg with many punctures (porous shell) would be adhesive during the filling or water-hardening process. As soon as the water-hardening process is complete, trout eggs are no longer adhesive.

Fertilization

The micropyle (Figure 8A), in its normal position in a newly-spawned egg, has the appearance of a small appendage resembling a minute cornucopia situated approximately at right angle to the outer shell and projecting inwardly. This small appendage commences to move to one side as soon as water entering through the pores starts mixing with the perivitelline fluid and fills the void between the outer shell and the yolk membrane. The micropyle is in its greatest open position at the time the egg is taken from the fish. As soon as it begins to move to one side, the opening decreases in size and continues to do so until the micropyle has moved into a position which completely seals off the opening to the outer shell. For this reason, the greatest chance for fertilization occurs immediately after the egg is taken. One can understand that the possibility of fertilizing the egg gets progressively poorer as the micropyle opening get smaller, and since the spermatozoon enters only through the micropyle, fertilization is impossible after sufficient time has elapsed to allow the micropyle to close.

The matter of whether the eggs or milt should be taken first is usually left to the individual. However, usually the female is spawned first, then the milt is added, and the eggs in the pan are stirred gently. If no broken eggs appear, another female is spawned into the pan and then another male. By using a different male each time, the chance of fertilization being upset by use of an impotent male is greatly reduced.

The number of eggs produced by females of the same age and strain varies considerably. The amount of sperm extruded from a male varies from a few drops to a teaspoonful. It has been stated that one drop of sperm will contain enough spermatozoa to fertilize 10,000 eggs. It is, of course, necessary for contact between sperm and eggs to occur; hence, the necessity of stirring the eggs and sperm together.

Since there is a limit in the time that both the eggs and the sperm remain viable, correct timing in the spawn taking operation is important. The length of time either eggs or sperm remain viable varies considerably and depends, perhaps, on several factors. Certainly, variety of fish and temperature are contributing factors. It is generally accepted that exposure of trout eggs or sperm to water for three minutes or more prior to fertilization will result in virtually a complete loss of viability.

In taking trout eggs, it is best to use a dark or black spawning pan, so it will be easier to see the milt in the pan. Should eggs be broken, the white albumen from the broken eggs is also more easily detected in a black pan. Any pan of suitable size can be darkened by painting the inside with asphaltum varnish.

Testing Fertilization

As early as 1902, glacial acetic acid was used to aid in determining whether salmon ova had been fertilized (Rutter, 1902). He noted that the embryos in eggs 24 hours or older, when placed in a 5 to 10% acetic acid solution, turned white within a few minutes. The dead ova are cleared by the acid, so that the stage of development at which the ova died may be determined. The further the embryo has developed, the easier it is to make a determination. In making this test for the first time, it is best to make a comparative test with unfertilized ova that have been kept in water during the same period. By noting the number of infertile ova in a sample, it is not difficult to determine, within reasonable limits, the percentage of fertilization in a lot of eggs.

Dry Method

The old subject of dry method versus wet method is still being debated and apparently both methods are equally successful, depending to a great extent on the individual spawntaker. Advocates of the dry method claim that the micropyle remains open longer, and they explain it in this way: water entering the egg through the pores and mixing with the perivitelline fluid causes the micropyle to close. In the dry method, the process is supposed to be delayed. With the micropyle remaining open longer, more time can be taken in adding the sperm, and the eggs from several females can be taken in the same pan. An experienced spawntaker who selects his fish carefully, making sure that both males and females are ripe, seldom is rushed for time, and unless broken eggs (albumen) enter the pan, he is able to spawn several females, just as when using the dry method.

Actually, it is difficult to take eggs without getting some water in the pan, because some will usually drip from the fish and the holder's glove.

Effect of Broken Eggs in Fertilization

When eggs are broken in the spawn-taking operation, the process of fertilization is greatly hampered and at times completely stopped. Broken eggs in the spawning pan will appear as a white, creamy substance somewhat resembling sperm. This is actually the albumen from the broken eggs and, unless it is washed off immediately, some of it will lodge over the micropyles and present the sper- matozoa from entering. Broken eggs probably contribute as much to poor fertili- zation as does any other factor. When albumen appears in the spawning pan, it should be washed off immediately, the sperm added, and the pan emptied of eggs before more are added.

Salt Solution in Spawn Taking

It has been demonstrated that when eggs are taken by inexperienced spawntak- ers or when eggs are extremely soft shelled, with many broken eggs resulting, fertilization can be improved by the use of a salt solution. A sufficient amount of the solution is added to the empty pan to fully cover the number of eggs to be taken, regardless of whether eggs from one or more females are taken. The salt solution holds the albumen from the broken eggs in solution and keeps the micropyles from becoming clogged. It also prevents agglutination of the sperm. An experienced spawntaker normally uses a salt solution only when eggs are soft and an excessive number of eggs are broken in stripping them from the fish. Various salt solutions have been developed and some latitude is allowable in their composition and application. A salt solution which has proven satisfactory for egg-taking purposes in California hatcheries can be prepared as follows: dissolve one ounce of common table salt in one gallon of water.

FIGURE 25—Artificially spawning a female trout by the two man method. *Photograph by J.H. Wales, 1959.*

Two Man Spawning Method

It is normal procedure in California to anesthetize the fish before spawning as described in the section on spawn taking facilities. There is much less danger of breaking eggs or injuring the fish if it does not struggle during this process.

In the two man method (Figure 25), one man holds the fish and the other spawns. The holder wears gloves on both hands and grasps the fish, male or female alike, by the caudal peduncle with the right hand and by the pectoral fins with the left hand. The egg pan is placed against a padded block on the spawning bench and the holder moves the fish over the spawning pan with his right hand resting on the padded block. The spawning pan is usually slightly recessed into the spawning bench, to prevent it from being knocked off the bench in the event a fish should struggle.

With the fish held tail down, so the ripe eggs will flow naturally toward the vent, the spawntaker, who stands on the opposite side of the bench, gently presses out the eggs with the thumb and forefinger, beginning pressure just forward of the vent. The hand is then moved forward toward the head of the fish and further gentle pressure is applied as necessary to assist the natural flow of eggs, until all that will come freely from the fish are obtained. Pressure should never be applied forward of the ventral fins, since even slight pressure applied over the heart and liver may injure the fish.

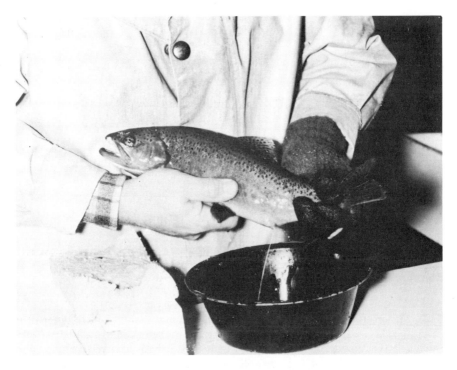

FIGURE 26—Taking sperm from a male trout by the single-handed method. *Photograph by J.H. Wales, 1959.*

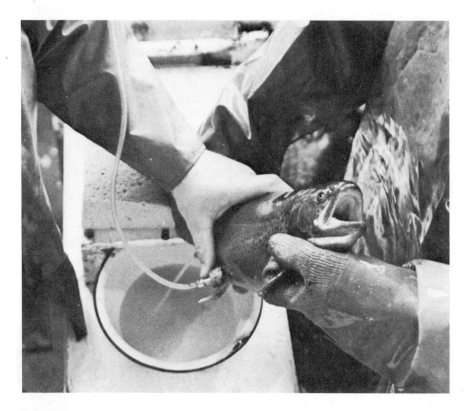

FIGURE 27—Air spawning a female trout. Hypodermic needle is inserted ½ to 1 inch into the body cavity near the pelvic fin. *Photograph by George Bruley, 1973.*

Do not attempt to strip the female entirely clean of eggs. The reason for this is obvious to those who have taken spawn, for it is in extracting the last hundred or so eggs that one is most apt to break shells and force from the fish eggs which may not have fully matured.

Single-Handed Spawning Method

In the single handed method, the spawner both holds and spawns the fish. The procedure is as follows:

The spawner, wearing a glove on the hand which will hold the tail, grasps the fish dorsal side up, holds it against his body with the ungloved hand, then strokes the eggs from the fish into the pan. (Figure 26.)

Air Spawning Method

In the air spawning method (Figure 27), one man holds the fish, after it has been anesthetized, in the same manner described in the two man method. Another man inserts a hypodermic needle ½ to 1 inch into the body cavity, depending upon the size of fish. The needle is generally inserted in the hollow under a pelvic fin for best results. The air pressure forced into the body cavity is about 2½ to 3 pounds. When the spawning has been completed, the needle is removed from the fish and the air in the fish is expelled by hand.

The equipment needed for this method of spawning is:

A small portable air compressor with tank
Oxygen valve with a low pressure gauge
Short length of surgical tubing
478 D Luer-Loc needle adapter
18G x 1½ regular point hypodermic needle
Air hose or copper tubing as required

Experiments have shown that a smaller number of eggs remain in the fish after air spawning than after the hand spawning method and in many cases, air spawning is faster. The eggs are cleaner and there are no broken shells. There is less chance of injury to the adult fish particularly by an inexperienced spawner. The air spawning method is now used extensively in California hatcheries.

Incision Spawning Method (Salmon)

There are five species of Pacific salmon, all of which die after spawning. They are:

(a) King salmon.
Other common names: chinook salmon, quinnat salmon, tye salmon, and spring salmon.
(b) Silver salmon.
Other common names: coho salmon, silversides, and hookbill salmon.
(c) Pink salmon.
Other common names: humpback salmon.
(d) Chum salmon.
Other common name: dog salmon.
(e) Sockeye salmon.
Other common names: red salmon and blueback salmon.

The nonanadromous form of the red salmon, introduced and established in certain California waters, is known as the kokanee, but elsewhere has also gone under the names of little redfish and silver trout.

Only two species, the king and silver salmon, are common in California (Figure 28). Since all Pacific salmon die after spawning, no advantage is derived by stripping the females. Instead, they are killed by a blow on the head and the incision spawning method is used. Milt, however, is stripped from the males, which may or may not be killed.

In the modern spawning facilities described earlier, the females are killed by a mechanical device called a "fish killer" (Figure 29). When the ripe females leave the sorting table, they are put down a chute head first on their backs. The fish is stopped when the head hits a metal door; a knife then automatically pierces the fishes' head. It is then released to proceed to a washing table where it is sprayed with water before spawning.

When taking salmon eggs by the incision method, at least three men are required. The holder grasps the female by the gill cover with a gloved hand and holds the head of the fish about waist high, with the tail hanging straight down and over the spawning pan. The spawntaker, from a sitting position, grasps the fish by the tail with one hand and holds the tail away from the spawning pan (Figure 30). This eliminates slime from driping into the pan. He then makes an incision in the side of the fish with a blunt tipped knife, starting just behind the

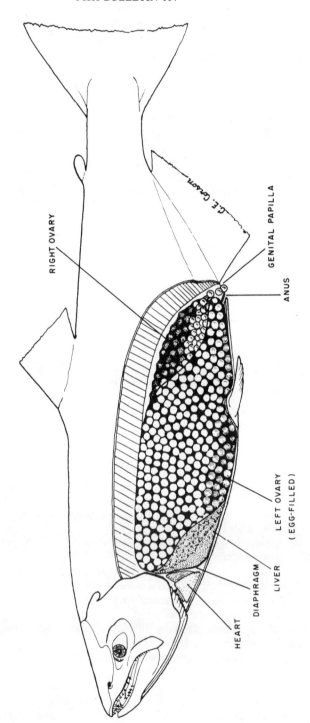

FIGURE 28—Ripe female king salmon.

A

B

FIGURE 29—**A**, ripe salmon to be spawned are taken from the sorting table and placed in the "fish killer" (shown in foreground) headfirst on their backs, as they slide down, the head hits a metal stop long enough for a knife to automatically pierce the head. **B**, view of automatic mechanism and knife on the "fish killer." *Photographs by George Bruley, 1973.*

pectoral fins and slightly to one side of the median ventral line and continuing to the genital papilla. Eggs begin to fall into the spawning pan when the incision reaches a point just above the ventral fins. The spawntaker then spreads his fingers apart and holds the body open so the ripe eggs will all fall into the spawning pan (Figure 31). No attempt should be made to force out any eggs still clinging to the ovaries. The person spawning the male sits to one side of the spawning pan and strips the milt from the male as required, making sure that each pan of eggs taken receives sufficient milt to fertilize them. Usually about a fluid ounce is used. Water is then added and the eggs in the pan are stirred.

FIGURE 30—Taking king salmon eggs by the incision method. *Photograph by Harold Wolf, 1958.*

Some of the salmon spawning facilities are arranged so one man can do the spawning. The spawner picks up the dead female from the washing table and hangs it on a hook by the gill cover opening. He proceeds to remove the eggs as described above, and then strips the milt from a dead male.

Bleeding of female salmon, to prevent excessive blood from entering the spawning pan, is not practiced in California state hatcheries. It is routine at most federal hatcheries. Blood, in its uncoagulated state, is an isotonic solution and, therefore, does not interfere with fertilization. Blood clots, however, prevent the sper-

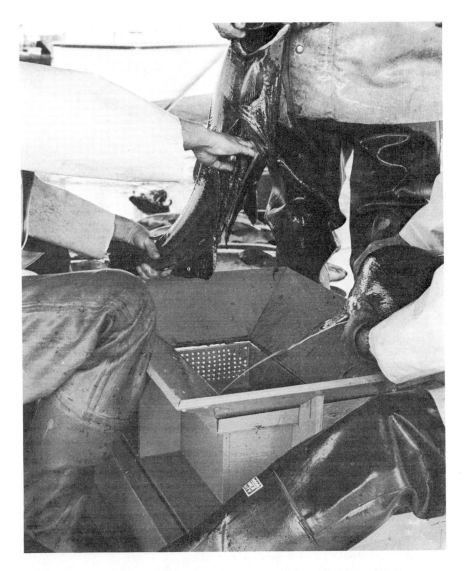

FIGURE 31—Incision is held open by spawntaker so that all ripe eggs will fall out while sperm from male fish is being added. *Photograph by Harold Wolf, 1958.*

matozoa from penetrating the eggs and, if present in excessive amount, may lower fertility. Bleeding of salmon females is done by slashing the caudal peduncle with a sharp instrument or by slitting the isthmus of the fish just anterior to the pectoral fins, thus severing the main artery and allowing the fish to bleed for several minutes before taking the eggs.

Delayed Fertilization

Eggs and sperm can be taken from fish and held separately for at least 24 hours before mixing to achieve good fertilization. The female fish must be dried with towels before spawning to keep any water from contaminating the eggs and ovarian fluid. The fish can be air spawned with the ovipositor placed in a tube so the eggs go directly into a plastic bag. The eggs can be taken from salmon by the incision method after the fish has been thoroughly dried.

The plastic bags containing the eggs should be sealed with most of the air removed and then placed in a styrofoam shipping case with crushed ice to maintain a temperature of about 38° F. The male fish must be dried thoroughly so the sperm can be extracted and put into plastic bags. These bags are also sealed with the air removed and placed in the case with the eggs. It is important that no water comes in contact with the eggs or sperm until the time of fertilization. It is preferable to put the eggs in a standard salt solution and add the sperm within 30 seconds. This procedure has been used with great success in California when the eggs are fertilized within 24 hours of spawning.

In selective breeding, crossbred fish can be developed by shipping eggs or sperm instead of adult fish. The shipment of unfertilized eggs is less complicated than fertilized green eggs because there is little loss due to eggs smothering.

Effect of Light on Trout and Salmon Eggs and Alevins

Nature provided that both trout and salmon eggs be shielded from light, especially from the direct rays of the sun. It has been proven that trout and salmon eggs are killed when exposed to direct sunlight for more than a few minutes at a time. In the modern egg incubator, in which trays are stacked one on top of another, the eggs are well shielded from light. In trough hatching, in which eggs are held in baskets, it is common practice to protect them from light by covering the baskets. The following incident may be cited. A trough cover, having a fairly round hole about 2 inches in diameter, was used. Sunlight shining through this hole killed the eggs which it reached, leaving a group of dead eggs about the size and shape of the hole in the cover.

Experiments with sockeye salmon eggs and alevins incubated in complete darkness showed decided physiological differences between them and eggs and alevins exposed diurnally to indirect daylight encountered in normal hatchery operations. Some of the observed major differences were as follows:

1. The light-exposed eggs hatched earlier than the eggs incubated in natural darkness.
2. The light-exposed alevins showed a decided negative reaction to light and constantly attempted to escape into a darker area, if available.
3. The light-exposed alevins were extremely active, while the alevins in the dark environment were relatively inactive.
4. The total mortality of eggs and alevins was greater when they were light-exposed.

5. Tests of both terminal swimming speed and swimming endurance showed that the fry which had been exposed to light during incubation and the post-hatching period were significantly weaker than those which had not been exposed to indirect daylight.
6. The light-exposed fry were smaller than the fry which had been produced in darkness and averaged one more vertebra.
7. The light-exposed alevins reached the "emerging fry stage" earlier than did the dark-exposed alevins.

The susceptibility of rainbow trout eggs to light is further indicated by experiments to test the possibility of shortening the incubation period with infrared ray lamps. In two experiments, first at Mt. Whitney Hatchery and later at Mt. Shasta Hatchery, all eggs exposed to the infrared light rays were killed. Harmful effects have also been reported when fluorescent lighting was used over incubating eggs. It seems doubtful that the small amount of ultraviolet rays emitted by fluorescent lights is harmful to trout and salmon eggs, and there is nothing to really substantiate this claim. However, to be on the safe side, it seems best to use incandescent lights. In any event, *always shield trout and salmon eggs and alevins from direct light rays.*

MEASURING EGGS

Newly taken or green eggs may be handled and measured as soon as they become water-hardened, and for roughly 48 hours afterwards, depending upon the water temperature. Green eggs should never be handled or moved unnecessarily during the water-hardening process, nor after they are 48 hours old or more, when development speeds up and the first stage of tenderness begins.

After the eye spots appear, trout and salmon eggs can be handled without any great danger of injury as long as they are not subjected to unusual shock, freezing, or dehydration.

Remember that eggs are sensitive to mechanical shock at all times and that reasonable care must be exercised in handling them.

Several methods for measuring and counting eggs have been developed. Although all have certain shortcomings, the California volumetric method, described here with two other well known methods, was adopted many years ago because of its accuracy and simplicity. Whatever method is chosen, for proper accuracy it is extremely important that the prescribed techniques and equipment be used.

California Volumetric Method

Equipment used in the volumetric method consists of the following (Figure 32): A, egg counting board; B, straight sided graduated beaker; C, 10 ounce measuring cup; D, 2 ounce measuring cup; E, bulb and egg picking tube; F, trimmed turkey quill.

In nearly all cases both green and eyed eggs are measured from a pail into the baskets or trays. In order to obtain an accurate count, the pail should contain just enough water to cover the eggs at all times. Surplus water in excess of that required to cover the eggs should be poured off. After a portion of the eggs has been measured out, it may become necessary to add a small amount of water. The water in the pail provides a cushion and prevents the eggs from being injured by the measuring equipment. It also allows them to flow readily into and out of the

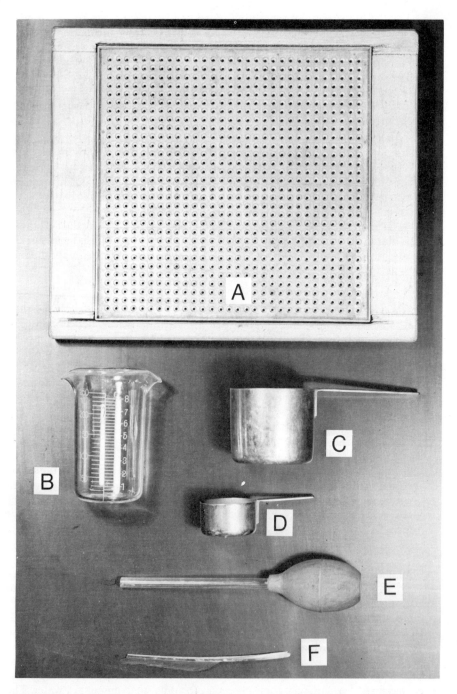

FIGURE 32—Egg measuring equipment. **A**, egg counting board; **B**, straight sided graduated beaker; **C**, 10 ounce measuring cup; **D**, 2 ounce measuring cup; **E**, bulb and egg picking pipette; **F**, trimmed turkey quill. *Photograph by J.H. Wales, 1958.*

measuring cup. Care should be taken never to jolt or jar a pail of eggs. In filling the measuring cup, it should be dipped into the eggs, completely submerged, and filled until it overflows. In emptying the cup, it should be tilted slightly to one side and the eggs allowed to flow from it. *Never allow the eggs to fall as they are being poured.*

The small 2 ounce cup is used as the unit measure in checking for count, and with it the number of eggs per liquid ounce is determined. It should be filled to overflowing. *Avoid compacting eggs into cup by shaking or pressure. Scrape off surplus eggs with an egg picking pipette.* The cup should be emptied into the egg-counting board and the trimmed turkey quill used to spread the eggs into the holes in the counting board. To assist in making the count, the counting board is marked off in squares of 25 holes each. The eggs in at least two or three small cups should be counted in making the check, the average being used in determining the number per ounce.

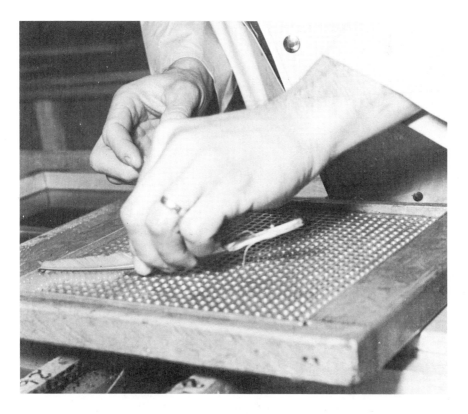

FIGURE 33—Eggs being counted on an egg counting board. *Photograph by George Bruley, 1973.*

Two ounces of eggs are counted on a counting board (Figure 33). Another rapid method of counting eggs is to place them on a wet, tightly stretched skaff net and use a pocket knife blade to move them from one pile to another in multiples of at least five eggs at a time.

The large cup holds exactly 10 liquid ounces and is to be used in measuring the eggs into the baskets or trays. Like the small cup, it should be filled to overflowing each time and the surplus eggs scraped off (Figure 34).

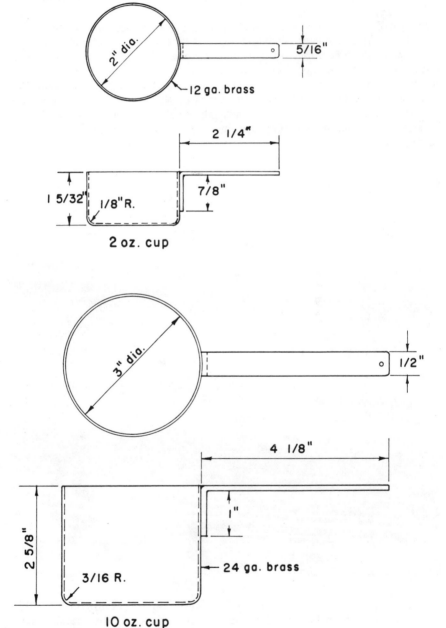

FIGURE 34—Egg measuring cups.

The graduated beaker, like the small and large cups, must have straight sides with the ounce graduations marked thereon. It is used in determining the number of ounces that remain at the end, when there is less than a full cup left.

Burrows Displacement Method

A displacement method has also been developed to enumerate salmonid eggs. This technique utilizes the volume of water displaced by the eggs. Two steps are involved in the technique: (1) measurements to determine the volume of water displaced by all the eggs, and (2) sampling to determine the number of eggs per fluid ounce of water displaced. This technique substitutes an accurately read water level, which has been displaced by the eggs, and provides a method for accurate measurements. The sampling technique, devised for the determination of the number of eggs per fluid ounce of water displaced, exactly duplicates the conditions of the volumetric method and thus automatically compensates for changes in the character of the eggs. Any greater accuracy, however, is attained at the expense of additional equipment and added time.

Von Bayer Method

The Von Bayer method employs a trout egg measuring trough (Figure 35). The trough is exactly 12 inches long, inside dimensions, having an angle of 45 degrees and a depth of approximately 2 inches. On the left side of the trough is a scale showing the number of eggs. On the right side is a scale indicating the

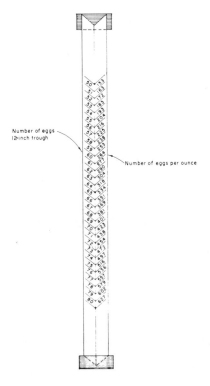

FIGURE 35—Von Bayer trout egg measuring trough.

FIGURE 36—Shocking eggs by the siphon method. *Photograph by Jerry Eskew, 1973.*

number of eggs per ounce. Eggs are hand counted as they are placed in a single row in the trough. If 60 eggs are placed in the trough and the Von Bayer scale is read opposite the figure 60, it will indicate 264 eggs per ounce. After the number of eggs per ounce has been determined from a sample, the rest are measured with a standard beaker or other acceptable utensil in fluid ounces, and the number of ounces is multiplied by the number of eggs per ounce.

SHOCKING OR ADDLING EGGS

The term shocking or addling trout and salmon eggs is applied to the process of turning the infertile eggs white so they can be separated from the fertile ones. Actually, this amounts to nothing more than agitating the eggs enough to rupture the yolk membrane in the infertile eggs, which causes them to turn white. Eggs can be shocked by stirring them in the basket with the bare hand. However, this may injure some of the good eggs by striking them against the sides and bottom of the basket. By far the safest and best way to shock eggs is to siphon them through a 4½ foot length of common garden hose from an egg basket in a hatchery trough to a pail on the floor (Figure 36). The pail should be perforated near the top edge, to allow the water to run off without carrying away the eggs. The distance from the end of the hose to the water surface in the pail will determine the degree of shock being given and can be varied until the desired results are obtained. *Trout and salmon eggs should not be shocked until the eye spots are clearly visible.*

PICKING EGGS

Pipette Method

Unless a fungicide is used to prevent dead eggs from fungusing and the fungus spreading to the fertile ones, it is necessary to remove the dead eggs. This is called egg picking.

Various methods have been developed to separate the white or dead eggs from the fertile ones. At one time, nearly all egg picking was done with a large pair of metal tweezers, a tedious process. A great improvement in technique was made when the pipette became universally adopted for picking eggs. It consists of a length of glass tube about eight inches long inserted into the bulb of an infant syringe. The inside diameter of the glass tube should be just large enough to pass the eggs. An experienced operator can pick eggs with a pipette quite rapidly. The operation, however, is quite tiring and can cause serious eye strain. In California, the pipette is used mostly for picking out small numbers of eggs preparatory to shipping. In making up pipettes for egg-picking work, the following materials are recommended:

Bulb, Goodrich, infant syringe, red, 2 ounces, No. 223, or 3 ounces, No. 224.

Tube, Pyrex, clear glass, 6 inch length, thickness of glass 1 mm. or more depending on inside diameter. The inside diameters of tubes commonly used are as follows:

    5 mm.—0.197 inches for eastern brook trout
    6 mm.—0.236 inches for small brown trout
    7 mm.—0.276 inches for brown and small rainbow trout
    8 mm.—0.315 inches for rainbow trout
    9 mm.—0.354 inches for steelhead
    10 mm.—0.394 inches for silver salmon
    11 mm.—0.443 inches for king salmon

Siphon Egg Picking Method

Various types of continuous flow, siphon type egg pickers have been developed, and great claims have been made for them. The continuous flow, siphon egg picker consists of the pipette with bulb, and a length of flexible hose reaching from the bulb at working height to a pail on the floor. The siphon is started and the egg picker moves the glass tube in the basket of eggs in much the same way as when using the pipette alone. The flow of water through the siphon hose is controlled by applying pressure to the syringe bulb, located between the end of the glass tube and the upper end of the siphon hose and held in the hand. Eggs can be picked faster with a siphon than with a pipette, because it is not necessary to empty out the bulb as is done with the pipette, since the eggs picked up are siphoned into the pail on the floor. A modification of the siphon egg picking device is the power egg picker described in the article, "Power Egg Picker" (R. J. McMullen, 1948).

Salt Flotation Method

The flotation method is one of the older methods still used to some extent. A salt box in which a standard egg basket will fit is used. The salt box is filled with water to nearly overflowing and common stock or table salt is added and stirred into the water until a solution of the proper strength is reached. The amount of salt to be added can best be determined by dipping up a sample of the solution in a glass container, preferably a quart fruit jar, and then dropping a few eggs into the jar, using both live and dead eggs. If the solution is right, the dead or bad eggs will float and the live or good eggs will slowly settle to the bottom. If both good and bad eggs float, the solution is too strong and should be diluted by adding water. If both good and bad eggs settle to the bottom, the solution is too weak and more salt should be stirred in. The margin at which the salt solution will separate the good eggs from the bad is quite narrow, so great care must be taken in the preparation of the solution. When a solution of the proper strength has been attained, the basket containing the eggs is set into the salt box, and after a moment or two the good eggs begin to settle to the bottom. The bad eggs floating on top are then skimmed off with a small scaph net. Care must be taken in skimming off the dead eggs and any turbulence created by the scaph net must be kept down, since any movement of the water will cause the good eggs to rise to the surface and mix with the bad. The eggs in a basket should not be over one inch deep for salting, regardless of the size of the eggs. Separation becomes more difficult with more than that amount.

For satisfactory salting, trout eggs should be well eyed. The further the embryo has developed, the more rapidly will the eggs settle in the salt solution. Eggs should be shocked at least 36 hours before salting, for good results. As the solution becomes diluted, more salt should be added. A salinometer may be used to determine the strength of the solution.

Once the strength of the solution has been determined, it is easy to maintain it by periodic testing. The optimum strength may vary with different lots, depending on the stage of development and the elapsed time between shocking and salting.

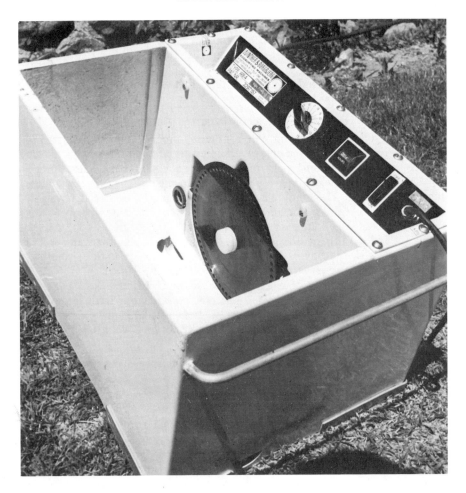

FIGURE 37—Mechanical egg sorter. Note the rotating disc with a row of holes around its circumference. A plastic guard that allows the eggs to reach the disc only at the bottom was removed so the disc would show in the picture. *Photograph by Jerry Eskew, 1973.*

Mechanical Egg Sorter

A machine has been developed that can sort about 100,000 eyed eggs per hour. The dimensions are about 16 inches × 20 inches × 13 inches deep. It has a magazine capacity of about 8 gallons (Figure 37). The machine can be set up in an area where water and electricity are available. The eggs are placed in the magazine and water washes them into the lower holes of a vertical rotating disc that has a row of holes around its circumference. The machine operates electronically on light sensitivity. As the disc rotates, the opaque, or dead, eggs are blown out of the disc by an air jet into a container. As the disc continues to rotate the clean, or good, eggs are blown out of the holes in the disc by a steady air current. These good eggs go into a container that is supplied with running water by a hose. Discs with different size holes are used for eggs of various sizes. The machine requires no attention other than reloading about every two hours and putting the good eggs into baskets or incubator trays. One man can operate three of these machines or carry on other duties if only one machine is used.

The machines can be rented at a nominal cost from Winther I. Borgbjerg, Hovedvejen 51, 8361 Hasselager, Denmark. The machines are equipped with an hour counter and rent is paid only for the time in operation. A similar type egg sorter will probably soon be available from a West Coast manufacturer.

Flush Treatment to Eliminate Egg Picking

Egg picking, once a tedious and time consuming process and still practiced at most hatcheries, has been found entirely unnecessary at others so long as fungus can be controlled. To eliminate egg picking, trout and salmon eggs are flushed once each day with a malachite green solution from the day they are taken until hatching commences. The solution is made by dissolving 1½ ounces, dry weight, of malachite green in 1 gallon of water. The flow in the hatchery trough is regulated to about 6 gallons per minute and then 3 liquid ounces of the stock solution are added to the upper end of the trough. As soon as the water in the trough has cleared, the flow is increased to normal. The strength of solution, water flow, and time required to flush are not extremely critical.

After the basket or tray of eggs has finished hatching, the dead eggs are disposed of. By previously having determined the percentage of fertilization, egg losses can be computed quite accurately. At locations where hatchery water flows through outside ponds, the flush treatment should not be done at feeding time.

FIGURE 38—Trout eggs hatching in egg basket. Note newly hatched alevins which have fallen through rectangular mesh into the trough. *Photograph by J.H. Wales, 1959.*

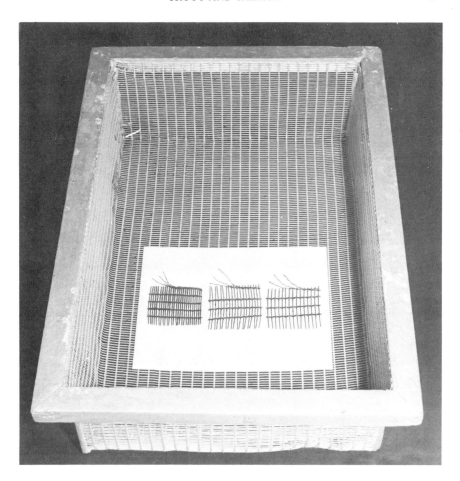

FIGURE 39—Fish egg-hatching basket made of triple warp mesh cloth. Note the rectangular openings. Swatches on white background show three mesh sizes. Left has 9 meshes per inch for brook trout and other small eggs from 400 to 700 per ounce. Center has 7 meshes per inch for rainbow and brown trout eggs 240 to 390 per ounce. Right has 6 meshes per inch for steelhead and silver salmon 120 to 380 per ounce. For king salmon eggs 60 to 90 per ounce, 5 meshes are required. *Photograph by J.H. Wales, 1958.*

TROUT EGG BASKETS

The popularity of egg incubators for incubating and hatching fish eggs has increased rapidly and the several advantages of the egg incubators are explained in another section. However, the egg basket is still used for hatching some eggs when hatchery troughs are available.

The standard California egg basket is 24 inches long, 14½ inches wide, and 6 inches deep. The outer frame to which the formed basket is attached is made of wood molding 1½ by ¾ inches. Egg basket mesh is a specially woven material with rectangular openings which prevent the round eggs from falling through. The newly-hatched alevins, on the other hand, being elongated and quite compressible, easily fall through the rectangles into the trough (Figure 38). When all of the fertile eggs have hatched, the basket is removed and the dead eggs disposed of.

Wire basket mesh is woven on order only and is available in rolls 27 inches wide and 100 feet long.

The material is known in the trade as mesh cloth, triple warp, and can be woven in a variety of sizes. The following description of mesh suitable for hatching rainbow and brown trout eggs will aid in describing basket mesh cloth.

Mesh cloth—triple warp, $\frac{5}{8}$ inch mesh, 0.035 inch diameter wire; fill, 7 inch mesh, 0.041 inch diameter wire.

Triple warp refers to the three 0.035 inch diameter wire warp which binds the mesh together (Figure 39). Five-eights inch mesh indicates the length of the opening. Fill 7 inch mesh indicates that there are seven wires 0.041 inches in diameter per inch, which determines the width of the opening. Mesh cloth is obtainable in galvanized, copper, brass, or aluminum material.

FIGURE 40—Wire-mesh fish egg basket suspended in hatchery trough. Note placement of metal division plates. Water flows over first plate and is forced downward by second plate, causing it to rise upward under egg basket for proper circulation. *Photograph by Harold Wolf, 1958.*

In basket hatching, up to five baskets can be placed in one trough and held until hatching time, and up to 150 ounces of trout eggs or 300 ounces of salmon eggs can be carried per basket. It is absolutely necessary, however, that division plates be placed ahead of each basket in such a way that the water flow will be directly underneath the basket and allowed to well up through the eggs for proper aeration (Figure 40).

FISH EGG INCUBATORS

Basically, fish egg incubators consist of a number of shallow trays stacked one on top of another and spaced apart by guiding strips, much as drawers in a dresser.

Some of the advantages of the egg incubator are the following: less space is required than for troughs; eggs are easily treated for fungus control; egg picking is eliminated; and, due to the small amount of water required, temperature may be controlled.

One of the first incubators was described by Nordqvist (1893). Since then several kinds have been described and used with varied success. They may be divided into two types: drip and the continuous or vertical flow. In the drip incubator, eggs are placed in the trays as soon as they are water-hardened and kept there until ready to hatch, at which time they are transferred to troughs. In the vertical flow incubator they are allowed to hatch there. The alevins remain in the trays until ready to feed. In general, the two types differ mainly in the way in which the water comes in contact with the egg. The vertical flow incubator is standard equipment in California because the alevins can be placed directly into tanks or concrete ponds to start feeding and it is not necessary to have hatchery troughs at the installations.

Vertical Flow Incubator

Incubators have several advantages over the conventional trough and basket method for incubating and hatching eggs. The flow of water through a 16 tray incubator varies from 3 to 10 gallons of water per minute. The flows are sometimes increased for eyed eggs and alevins as they require more oxygen than the green eggs. Larger flows are used when a tray is loaded with an extra large number of eggs. The average flow generally used in a 16 tray incubator is 5 gallons per minute. This small amount of water permits recirculating and either heating or cooling the water to speed up or delay development, whichever may be desirable.

To control the temperature, a portion of the recirculated water passes through a chiller for cooling water or a gas fired water heater and a filter (Figure 41). Another method of heating the recirculated water is to drain the water from the incubator into a well in the floor in a heated room. The room heat usually keeps the water at a very desirable temperature as it is recirculated.

In the vertical flow incubator, the trays actually consist of two compartments: first, the basket with cover in which the eggs are held, and second, the tray in which the basket rests. The trays are stacked one on top of another (Figure 42). Water flows from a tray to the tray immediately below, until it is drawn off at the bottom. The water is introduced to the bottom of each tray in such a manner that it wells up through the basket containing the eggs. This provides sufficient aeration for their development. In the vertical flow incubator, the eggs are al-

FIGURE 41—Equipment for use in controlling temperature of water supply for incubators. **A**, a chiller for cooling water; **B**, a gas-fired water heater; and **C**, a water filter. *Photographs **A**, by George Bruley, 1973; **B** and **C**, by Harold Wolf, 1958.*

A

B

C

FIGURE 42—**A,** Views of a battery of 16 tray vertical flow fiberglass incubators; **B,** measuring salmon eggs into an incubator tray; and **C,** alevins in an incubator tray. *Photographs by George Bruley, 1973.*

lowed to hatch and the alevins remain in the trays until ready to feed. For this reason, the trays must be covered to prevent the newly hatched fish from escaping.

Fungus is controlled with malachite green flush treatments. A stock solution of malachite green is made by dissolving $1\frac{1}{2}$ ounces of dry malachite green in one gallon of water. The flush treatment consists of pouring 3 ounces of the stock solution into the top tray of the incubator stack when the water flow is 5 or 6 gallons per minute. This flush treatment is generally continued on a daily basis. Most hatcheries discontinue the treatment when the eggs start to hatch, while others continue the treatment until the alevins are ready to feed and are removed from the incubator.

When the water is recirculated through the incubator, the malachite green flush treatment is done only two or three times per week. The treated water must be wasted and the system filled with fresh water which will probably change the temperature in the incubator for a few hours.

When the fish have absorbed their yolk sac and are ready to feed, the tray units can be moved to troughs, tanks, or ponds, placed on the bottom, cover screens removed, and the fish allowed to swim out. Dead eggs and deformed fish will remain in the egg tray and can be counted and then discarded. This accounts for the entire egg and fry loss to the feeding stage. There is no need to remove the dead eggs earlier as the malachite green controls the fungus. When eyed eggs are to be shipped, the dead eggs should be removed before packing them so fungus will not develop while the eggs are in transit.

The number of eggs put into each incubator tray varies with conditions and facilities available. Salmon eggs have been hatched successfully with 160 ounces or 12,000 fry per tray, although the California average is about 110 ounces or 8,000 fry per tray. Rainbow trout eggs have been hatched successfully with 100 ounces or 45,000 fry per tray, though the California average is about 70 ounces or 25,000 fry.

Heath Tecna-Plastics, 19819 - 84th Avenue South, Kent, Washington 98031 manufactures incubators made entirely of fiberglass.

PACKING EGGS FOR SHIPMENT

Green Eggs

Green eggs may be shipped for a period of up to 48 hours after being taken. Normally, when green eggs are transported only a short distance they are placed in 10 gallon milk cans filled with water. This method is quite satisfactory so long as severe jolting and rise in temperature are prevented. Care must be taken not to place too many eggs in a can. It must be remembered that trout and salmon eggs increase in size by about 20% during the water-hardening process, and if too many eggs are placed in a can the eggs near the bottom may be killed by the resulting pressure from the eggs above. A simple way to prevent excessive pressure on the eggs in the bottom portion of the can is to measure out eggs in amounts of 50 ounces each and place them on pieces of mosquito netting about 18 inches square. The four corners of the netting are then picked up and tied with a string. The eggs can then be suspended at different depths in the can by varying the length of the suspending string.

Green eggs that have not been fertilized can be sealed in plastic bags and transported in a styrofoam ice chest with crushed ice to maintain a temperature of about 38° F. as described in the section on delayed fertilization.

Eyed Eggs

Eyed salmonid eggs have been shipped to all corners of the world in a variety of shipping containers. In selecting shipping containers, one should bear in mind that eyed eggs must be kept cold to prevent their hatching en route.

Good judgment is necessary on the part of the hatcheryman when making egg shipments. Eggs must be kept moist to prevent dehydration and containers should be light to keep shipping charges at a minimum and should protect the eggs as much as possible against sudden shock. Eggs should not be so far advanced that they will hatch en route and the resulting alevins lost. The temperature in a shipping case is usually higher than that of most hatchery waters and fungus will grow very rapidly on dead eggs if they are left in the shipment. For shipping purposes, eggs should be sufficiently eyed to permit shocking, so that the infertile eggs can be removed.

The term eyed eggs is used rather loosely in hatchery practice and is usually applied to the eggs any time after the eye spots are visible to the unaided eye.

A lightweight egg case is used by California for shipping eyed eggs by truck or air (Figure 43). The cases are cardboard cartons 15 inches square by 23 inches in height, number V-3C carton, available from Crown Zellerbach Paper Company. The trays are moulded at the hatcheries using Polylite which is a polyurethane foam. A case consists of one ice tray, four egg trays, and the bottom tray to catch the water from melting ice. The trays are 14¼ inches square by 3½ inches deep. The ice tray has a ⅛ or ¼ inch screen molded into the bottom of the tray while the egg trays have aluminium window screen usually 14 x 18 mesh and the water tray has a piece of plastic molded in the bottom to catch the water.

These trays can be packed with 200 to 250 ounces of eggs. It is optional whether the eggs are packed in a piece of cheesecloth or just loose in the tray. The principle advantage to packing the eggs in cheesecloth is that they can be removed from the trays easier and placed into a bucket. A second advantage is that the cheesecloth retains some additional moisture around the eggs on long trips.

The trays are packed inside a plastic bag in the carton—water tray on the bottom, four egg trays, and the ice tray loaded with ice on top. The top of the bag is folded loosely over the ice tray and the carton closed and sealed. Air holes are cut in the top of the case. Care should be exercised to see that the carton remains dry while being packed and that excess water is drained from the eggs before packing the trays in the case. These cases properly packed will not leak and can be shipped by air.

The egg cases can be made for a minimal cost of 10 to 12 dollars and are generally used only once as a means of preventing the spread of any disease that might be present at the hatcheries involved.

Use of Wescodyne

As a further disease prevention measure all eggs received at a hatchery are treated with Wescodyne as they are removed from the egg case. Wescodyne rapidly destroys bacteria, viruses, molds, and fungi. Toxicity to man is exceptionally low.

A

B

FIGURE 43—**A**, lightweight egg shipping case; and **B**, tray packed with eggs which is placed in the cardboard carton inside a leak proof plastic liner. *Photographs by Jerry Eskew, 1973.*

Each lot of eggs, upon arrival, should be treated with a 1:300 solution of Wescodyne. Either green or eyed eggs should be completely submersed in the disinfecting solution for 10 minutes and then rinsed in fresh water. A 1:300 solution is formulated by using 12.6 ml of Wescodyne per gallon of water (4¼ fluid ounces in 10 gallons). The volume of eggs to disinfecting solution should be about 1 to 5 (25 ounces of eggs per gallon of disinfectant). The resulting solution may be used until the amber color fades to a light yellow. It should then be discarded and replaced with a fresh one. This formulation should also contain sufficient oxygen at all times.

In extremely soft water hatcheries; i.e., those where the alkalinity is less than 35 ppm, a buffer should be added with the Wescodyne to prevent a drastic drop in the pH to under 6. If in doubt, check the pH prior to treatment. Sodium bicarbonate is recommended as a buffer at the rate of 0.5 gram per gallon of disinfecting solution.

Wescodyne at a concentration of 1:20,000 is harmful to fish; therefore, care must be taken during disposal. *Do not* introduce it back into the hatchery system or into any septic systems dependent on bacterial decomposition.

Wescodyne may be obtained from: West Chemical Products, Inc.
8125 - 36th Avenue
Sacramento, California 95824

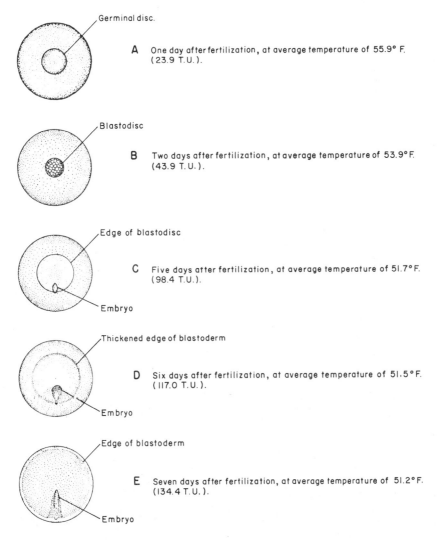

Germinal disc.

A   One day after fertilization, at average temperature of 55.9° F. (23.9 T.U.).

Blastodisc

B   Two days after fertilization, at average temperature of 53.9° F. (43.9 T.U.).

Edge of blastodisc

C   Five days after fertilization, at average temperature of 51.7° F. (98.4 T.U.).

Embryo

Thickened edge of blastoderm

D   Six days after fertilization, at average temperature of 51.5° F. (117.0 T.U.).

Embryo

Edge of blastoderm

E   Seven days after fertilization, at average temperature of 51.2° F. (134.4 T.U.).

Embryo

FIGURE 44—Development of steelhead trout eggs.

DEVELOPMENT OF TROUT AND SALMON EGGS

Trout and salmon eggs undergo a continuous developmental change from the time they are taken until they hatch. The rate of development is dependent on water temperature. During this period there are several changes or stages which are important for fish culturists to recognize. Important stages may be briefly summed up as follows:

1. *Fertilization.* This takes place within seconds after the eggs are taken and is dependent on several factors, such as degree of ripeness of the male and female fish, viability of both sperm and ova, and technique of the spawntaker.

2. *Water hardening.* This is the period during which the egg absorbs water and becomes firm and slick. From the time the egg becomes water-hardened and for a period up to 48 hours, depending on water temperature, newly taken trout and salmon eggs may be measured and shipped if this is carefully done.

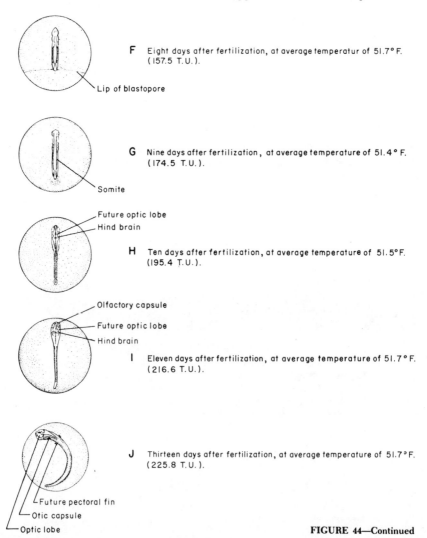

F   Eight days after fertilization, at average temperatur of 51.7° F. (157.5 T. U.).

Lip of blastopore

G   Nine days after fertilization, at average temperature of 51.4° F. (174.5 T.U.).

Somite

Future optic lobe
Hind brain

H   Ten days after fertilization, at average temperature of 51.5°F. (195.4 T.U.).

Olfactory capsule

Future optic lobe
Hind brain

I   Eleven days after fertilization, at average temperature of 51.7° F. (216.6 T.U.).

J   Thirteen days after fertilization, at average temperature of 51.7°F. (225.8 T.U.).

Future pectoral fin
Otic capsule
Optic lobe

FIGURE 44—Continued

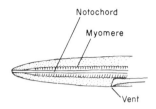

K   Fourteen days after fertilization,
    at average temperature of 51.5°F.
    (273.2 T.U.).

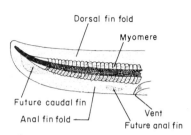

L   Sixteen days after fertilization,
    at average temperature of 51.7° F.
    (315.9 T.U.).

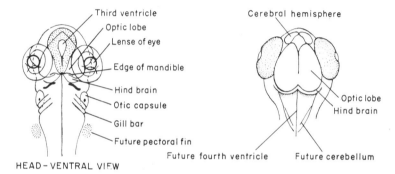

HEAD – VENTRAL VIEW

M   Sixteen days after fertilization,
    at average temperature of 51.7°F.
    (315.9 T.U.).

N   Eighteen days after fertilization,
    at average temperature of 51.8° F.
    (357.4 T.U.).

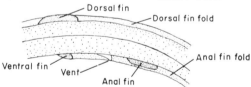

O   Twenty-six days after fertilization, at average temperature of 51.2°F. (500.4 T.U.).

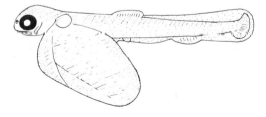

P   Hatched

FIGURE 44—Continued

3. *Tender period.* Trout and salmon eggs become progressively more tender during a period extending roughly from 48 hours after water hardening until eyed. The extreme critical period for steelhead trout eggs is entered into on the seventh day in water having a temperature of 51 °F. (Figure 44E) and extends through the ninth day (Figure 44G), at which time the blastopore is completely closed. It is common practice to work or pick eggs from the second day after taking until the critical period is reached. Then they are not touched until the critical period has passed. This is usually referred to as the period during which the eggs are closed. Even though the critical period for steelhead eggs has passed on the ninth day at 51° F., the eggs remain tender until the 16th day, when the eyes are sufficiently pigmented to be visible.

4. *Eyed stage.* As the term implies, this is the stage from the time the eye spots become visible until the egg hatches. During the eyed stage, eggs are usually addled, cleaned up, remeasured, shipped, or set out for hatching. The oxygen requirement for trout and salmon eggs increases as the embryo develops and smothering of eyed eggs in baskets or incubator trays must be prevented.

Temperature Units

For many years temperature units were used by hatcherymen to estimate the length of time required for trout eggs to develop, and it was believed that the total number of units was constant for any species regardless of the average temperature throughout the incubation period.

One temperature unit (T.U.) equals one degree Fahrenheit above freezing (32° F) for a period of 24 hours. For example, if the average water temperature for the first day of incubation was 55.9° F it would constitute 23.9 T.U. (55.9° F minus 32° F). If, on the second day, the temperature averaged 52° F, it would add 20 more T.U., or a total of 43.9 to the end of the second day.

Temperature units vary considerably at different water temperatures (Table 4). It clearly demonstrates thay they are not a safe measure for the trout culturist. They are included here in the tables and drawings only for the purpose of illustration.

TABLE 4

Number of Days and Temperature Units Required for Trout Eggs to Hatch *

Water temperature in degrees F.	Rainbow		Brown		Brook		Lake	
	Number days to hatch	Temperature units	Number days to hatch	Temperature units	Number days to hatch	Temperature units	Number days to hatch	Temperature units
35	--	--	156	468	144	432	162	486
40	80	640	100	800	103	824	108	864
45	48	624	64	832	68	884	72	936
50	31	558	41	738	44	799	49	882
55	24	552	--	--	35	805	--	--
60	19	532	--	--	--	--	--	--

* Spaces without figures indicate incomplete data rather than a proved incapability of eggs to hatch at those temperatures.

In any species of trout, the incubation period or average hatching time of the eggs is not immutably fixed for a given temperature but may vary as much as 6 days between egg lots taken from different parent fish.

The development of sea-run steelhead eggs takes 30 days at 51° F. (Table 5; Figures 44A through 44P). Other strains of the rainbow group have a similar pattern.

TABLE 5

Record of Daily Steelhead Egg Samples

In the following descriptions a few of the more important facts have been given for each of the daily egg samples. These samples were taken at the same time each day and the temperature units listed refer to the total accumulated up to the time that particular sample was taken.

Number of days since fertilization at average temperature of 51 degrees F.	Temperature units	
1	23.9	Blastodisc without cell division (Figure 44A).
2	43.9	Blastodisc has rounded up and is slightly smaller in diameter; it is composed of three layers of cells (Figure 44B).
3	62.2	Blastodisc slightly larger; cells about $\frac{1}{2}$ the size of those in sample 2.
4	81.1	Blastodisc slightly larger; cells about $\frac{2}{3}$ the size of those in sample 3.
5	98.4	Blastodisc slightly larger; edge thickened; embryo started to form (Figure 44C).
6	117.0	Thickened edge of blastoderm is narrower; embryo with thickenings (Figure 44D).
7	134.4	Blastoderm covering about $\frac{2}{3}$ of the yolk surface; embryo much lengthened and thickened, strongly contrasted against the yolk; eggs are now tender and should not be handled roughly (Figure 44E).
8	157.5	Optic vesicles forming; body with 23 somites; blastoderm covering about $\frac{3}{4}$ of the yolk (Figure 44F).
9	174.5	Blastopore closed; optic vesicles have invaginated; notochord continuous with brain, end of critical period (Figure 44G).
10	195.4	Eye lenses well developed; 32 somites present; otic vesicles starting to form; notochord continuous with brain (Figure 44H).
11	216.6	Cerebral hemispheres evident; segmentation of hind brain begun; 46 somites present; pectoral fins starting to form; notochord ending just anterior to otic capsules; the caudal region apt to be twisted (Figure 44I).
12	236.8	Optic lobes starting to grow over interior-posterior margin of eyes; lenses not quite broken away from ectoderm; notochord ending just posterior to anterior end of otic capsules; 49 somites present (the last 5 have not become myomeres); the embryo covers about $\frac{3}{4}$ circumference of egg.
13	255.8	Lenses have broken away from the ectoderm; 60 somites present (all have become myomeres); 3 gill buds present (Figure 44J).
14	273.2	Fin folds about half as wide as body; tail still homocercal; mouth and gill cartilages faintly evident (Figure 44K).
15	297.8	Tail becoming heterocercal; eyes becoming pigmented.
16	315.9	Four gill buds evident at sides of throat; hereafter there are but slight changes in the brain; the hind brain apparently decreases and the cerebral hemispheres enlarge; anal fin first noticeable here; the eggs are past the tender period and can be handled more roughly (Figures 44L and M).
17	336.6	No apparent changes except in size.
18	357.4	Dorsal fin begins here as a faint thickening; nostrils first evident; gill slits on ventral side extending almost to midline (Figure 44N).
19	377.7	Opercle grown back over first gill bar; traces of melanophores on head and body.
20	395.3	Gill bars have become definite arches; mandible has become pointed up between eyes but still far from maxillary.
21	412.0	Increase in size.
22	430.8	Ventral fins now evident.
23	448.0	Cerebral hemispheres about $\frac{2}{3}$ length of optic lobes.
24	466.6	No apparent changes.
25	485.4	Point of opercle extending back to 4th gill arch, only the distal ends of last 3 exposed.
26	500.4	No apparent changes (Figure 44O).
27	518.7	Mandible apparently stops its forward growth at this stage.
28	535.9	Dorsal and anal fin rays first visible at this age; point of opercle entirely covering first 2 gill arches; hatching begins.
29	556.1	Length of embryo equal to circumference of egg; anal fin shows about 8 rays and dorsal fin shows about 6.
30	573.9	Hatching completed (Figure 44P).

HATCHERY TROUGHS

The standard California hatchery trough is 16 feet long, 16 inches wide, and 7½ inches deep, inside measurements, with a gradient of one inch in 16 feet, and when operated with a 5 inch outlet plug contains roughly 64 gallons of water. The flow through the troughs varies among hatcheries and depends on the temperature and number of eggs or fish being carried in the troughs. On the average, troughs in California hatcheries, regardless of whether they are installed singly or in tandem, receive from 12 to 15 gallons of water per minute. It is intended that the oxygen content be maintained at not less than 7 ppm.

At one time, all troughs were made of wood, either cedar or redwood being used. In recent years, the trend has been toward metal (aluminum) troughs (Figure 45). Advantages of aluminum troughs are that the normal tasks such as scrubbing, sanding, and painting are practically eliminated. There is less chance for leakage and the smooth nonporous surface does not provide as great a haven for disease organisms as does the old wooden trough. A wide variety of aluminum alloys, differing in chemical composition and physical properties, have been developed. The various aluminum alloys are designated by numbers, such as 3514-½-H, 53ST, etc. A complete nomenclature may be obtained from any of the several large manufacturers of aluminum alloys, and with the aid of this nomenclature the various alloys can be identified.

FIGURE 45—Aluminum hatchery troughs. Center water supply permits unobstructed aisle between troughs from one side of building to the other. *Photograph by Harold Wolf, 1958.*

It is very important for anyone contemplating the installation of aluminum troughs to first obtain a complete chemical analysis of the water at the location where the troughs are to be installed. This analysis should be made available to a reliable aluminum manufacturing company, which will usually furnish competent engineering assistance in selecting the proper alloy.

Trough Carrying Capacity

It is generally agreed that the oxygen requirements for salmon and trout are nearly the same for all species, with rainbow possibly requiring the highest amount, brown trout the next highest, and eastern brook the lowest. Oxygen requirements are increased by activity, feeding, and rise in temperature. While it is desirable to maintain an oxygen level of 7 ppm for holding salmon and trout, the absolute minimum for this purpose is 5 ppm. It has been shown that at 45° F. temperature a yearling rainbow will consume 3 cc of oxygen per hour, while at 68° F. consumption will increase to 12 cc per hour, an increase of 300%. The increase in oxygen use, associated with rising temperature, crowding, feeding, weighing, and activity, must be considered when determining the number of fish to place in hatchery troughs and ponds (Table 6).

TABLE 6

Trough Carrying Capacity in Ounces of Trout Per Gallon Per Minute of Water Flow. Inlet Water Saturated. Outlet Water Oxygen Content 5 p.p.m.*

Temperature in degrees F.	Elevation in feet										
	0	1,000	2,000	3,000	4,000	5,000	6,000	7,000	8,000	9,000	10,000
	Ounces of trout										
45	400	380	350	330	310	285	265	240	225	210	190
50	210	195	180	170	160	145	135	120	110	100	95
55	132	125	116	107	100	93	83	76	69	62	55
60	93	85	80	72	67	62	56	50	45	39	34
65	66	61	57	51	46	42	37	33	30	25	21
70	52	48	44	39	36	31	28	24	22	18	14

* At lower temperatures and altitudes, actual crowding of fish would become a limiting factor, rather than available oxygen. Thus, at 45 degrees F. and sea level, 20 gallons per minute would supply 8,000 ounces with sufficient oxygen, but that weight of fish would require more than a standard trough to prevent overcrowding.

There are minimum water requirements in a hatchery trough (Table 7). Data for troughs are not applicable to rectangular or circular ponds, in which reaeration from the surface is large in comparison to oxygen directly available from the water supply.

Even though it is not general practice in California hatcheries to install hatchery trough aerators, the oxygen level in troughs can be considerably increased by their use. Reference is made to "Hatchery Trough Aerators," by Paul A. Shaw (1936).

TABLE 7

Minimum Water Requirements in Hatchery Troughs in Gallons Per Minute for Each 1,000 Ounces of Trout *

Temperature in degrees F.	Elevation in feet										
	0	1,000	2,000	3,000	4,000	5,000	6,000	7,000	8,000	9,000	10,000
	Gallons per minute										
45	2.5	2.7	2.9	3.0	3.2	3.5	3.8	4.2	4.5	4.8	5.3
50	4.8	5.4	5.6	5.9	6.2	6.9	7.4	8.3	9.1	10	11
55	7.6	8.0	8.6	9.3	10	11	12	13	15	16	18
60	11	12	13	14	15	16	18	20	22	26	29
65	15	16	18	20	22	24	27	30	33	40	48
70	19	21	23	26	28	32	36	41	45	55	71

* Values above 10 are shown only to the nearest gallon. For tandem troughs with good aerators between upper and lower, the water requirements can be lowered about 10%.

TROUT AND SALMON PONDS

Pools or ponds for rearing trout and salmon have over the years been as varied in design as French fashions. One can understand that in large, deep, still pools, fish rest most of the time and food conversion is normally better than in narrow, swift ponds, where a good portion of a fish's energy is used in maintaining its position in the pond. The ideal pond is one that can be operated rather deep and with little current most of the time, but can be readily converted to a shallow, swift pond when necessary. Deep, still pools have the disadvantage of not lending themselves to flush or prophylactic treatments, whereas long, shallow, raceway type ponds are ideal for this purpose. Very often the type of pool selected depends on several factors, such as amount of water available, surrounding terrain, and accessibility. In fish culture as it is practiced today, ponds must lend themselves to ease in flush treatment for prevention of disease, automatic grading, accessibility to mechanical loading and feeding equipment, and simplicity in overall management. In California, concrete raceway ponds have proven to be the most practical.

Rearing Pond Capacities

The carrying capacities of rearing ponds depend, to a large extent, on conditions existing at the various hatcheries. Capacities must be established at the hatchery itself to be of benefit in planning the station program. Some of the factors which influence or determine the carrying capacities of rearing ponds are: (1) water quality, (2) water temperature, (3) water volume, (4) rate of flow, (5) rate of change, (6) reuse of water, (7) degree of pollution of water supply, (8) kind of fish held, (9) size of fish held, (10) frequency of grading and thinning, and (11) diseases encountered.

The best way to determine the proper holding capacities of rearing ponds at a particular installation is to examine the results of several seasons of production for which accurate records are available regarding numbers and weights of fish held, growth, food conversion, types of feed used, incidence of disease, and mortality—all plotted against the basic factors, such as water volume and temperature, which might influence holding capacities.

Circular Ponds

The advantages of the circular pond are: less water required; nearly a uniform pattern of water circulation throughout the pool, with fish more evenly distributed instead of congregating at the head end; center outlet, with circular motion of water producing a self-cleaning effect. Even though less water is required, sufficient head or pressure must be available to force and maintain the water in a circular motion. In operating a circular pond, advantage can be taken of an interesting phenomenon. Since the vortex of a whirlpool in the northern hemisphere always rotates in a counterclockwise direction, the water in a circular pond should also rotate in a counterclockwise direction.

To take advantage of the self-cleaning effect of the circular pond, it is necessary that the proper type screen and outlet pipe be used. In effect, the self-cleaning screen consists of a sleeve larger than the center outlet or standpipe. This sleeve fits over the outlet pipe and projects above the surface of the water. The sleeve may have a series of slots or perforations near the bottom which act as an outlet screen for small fish or may be in the form of a narrow opening between the pool bottom and the lower end of the sleeve for larger fish. This opening must be adjusted according to the size of the fish in the pool. Waste materials are drawn through the opening by the outflowing water.

While carrying capacities will vary among installations, circular tanks 14 feet in diameter, with a water depth of 30 inches and an inflow of 50 gallons per minute, will safely carry 400 pounds of trout at water temperatures up to 60° F. With a comparatively small amount of water used in a circular pool, the rate of water exchange is quite slow and extreme care must be taken in feeding.

FIGURE 46—Small fiberglass circular tanks can be placed in groups so one Allen feeder can supply dry feed to two tanks. *Photograph by George Bruley, 1973.*

Circular ponds do not lend themselves well to flush treatment for disease control nor to mechanical fish loading or self-grading devices. The advantages of the circular pond for yearling and larger trout are not as great as thought when circular ponds first began to appear. The circular pond, however, is well adapted to rearing trout ranging from newly hatched fry to fish of subcatchable size.

There are many advantages to circular tanks in holding a small number of fish of varying ages and sizes for selective breeding, feed experiments, and many types of research projects. In these cases it is desirable to have a separate water supply for each experiment. Generally there are only a small number of fish involved in each experiment and a large number of small tanks is very practical.

Small circular fish tanks constructed of fiberglass reinforced polyester plastic are available in sizes of 3, 4, 5, and 6 feet inside diameter. Depth of the 3 foot tank is 26 inches; the 4 foot tank is 30 inches deep; while 5 and 6 foot tanks are 36 inches (Figure 46). The drain pipe and screen devices are similar to the 14 foot diameter tank described earlier.

These tanks are available from:

Helnic Manufacturing Co.
3011 Q Street
North Highlands, California

Edo Western Corporation
2845 South 2nd West
Salt Lake City, Utah

Heath-Tecna Corporation
Plastics Division
19819 - 84th Avenue South
Kent, Washington 98031

Raceway Ponds

The earthfill raceway type rearing pond with concrete cross dams was at one time the most widely used pond in California hatcheries. They were constructed 100 feet long, 4 feet deep, 10 feet wide at the bottom and 30 feet wide at the top, with sides sloped 2½ to 1. The gradient was 6 inches in 100 feet. They were usually built in a series of four to eight ponds long with a roadway along one side of each series.

These earthfill ponds require considerable maintenance as weeds and plants grow on the banks and protrude into the pond. The pond banks erode making the banks between ponds and roadways quite narrow unless they are reshaped annually. Flush treatments for disease control in ponds with irregular widths is not satisfactory as they have poor circulation.

Concrete raceway type rearing ponds have been built in hatcheries constructed since 1964 and have become the standard type rearing ponds in California, replacing earth type ponds.

The standard concrete raceway type rearing pond is 100 feet long, 42 inches deep, and 10 feet wide inside measurements, with a 6 inch gradient per 100 feet. The top of the pond wall is generally 18 inches above ground level. The depth of water in the pond at the lower end varies from 24 to 30 inches depending on various conditions at a particular hatchery; i.e., size and number of fish per pond, water flow, water temperature, diseases present, and fishes increased desire to

FIGURE 47—Fresh-Flow aerators at the upper end of pond number six in six-pond series. Each sprays about 1 cfs of water and increases the oxygen about 2 ppm. Note wires over the ponds to keep out sea gulls. *Photograph by George Bruley, 1973.*

jump out of the ponds at certain times of the year. The flow of water in a series of six ponds is usually about 3 cfs. Pond series are constructed from four to ten ponds in length but general practice is to build only six ponds in a series. A water flow of 5 cfs should be available for a series ten ponds in length. In a series of six, and particularly ten ponds the water should be pumped through an aerator midway in the series so the lower ponds can operate at full capacity as was previously mentioned.

About 2 ppm oxygen can be added to a series of ponds by operating a Fresh-Flow aerator in the upper end of pond five of a six pond series, thereby increasing the production in ponds five and six. This type aerator is operated by a 1 horsepower motor and sprays about 1 cfs of water into the pond (Figure 47).

Two series of ponds are normally built side by side with a common wall between them. There is a roadway between each pair of ponds. The roadways give access to all ponds for the mechanical pellet blower to feed the fish, the fish pump, and tank trucks. A battery of ponds as described above can be constructed in a much smaller area than earthen ponds (Figure 5).

Keyways in the concrete walls at 100 foot intervals provide for screens and check boards. At these points metal posts with keyways are placed in recessed holes in the middle of the concrete floor. Thus two screens about 5 feet long are used instead of one 10 foot screen (Figure 48). The water drops 6 inches over the checkboards at the end of each pond which provides some aeration. It is advantageous to have keyways in the concrete walls at 25 foot intervals in the upper ponds of a series so small fish can be crowded together for proper feeding.

FIGURE 48—Concrete rearing ponds are 10 feet in width with keyways in the concrete walls for screens and checkboards. Note metal removable post that fits in a pocket in the concrete floor so two 5 foot screens can be used instead of one long screen. *Photograph by George Bruley, 1973.*

Fry are taken from the incubator trays when they reach swimup stage to start feeding. The number of fry to stock an entire series of ponds can be placed in the upper end of the series and kept crowded until they are feeding well. They are then released into additional ponds in the series or moved to other ponds as required. It is always preferable to keep the small fish in the upper end of the series and the larger fish below. With proper management planning, it is possible to keep most of the ponds full at all times with fish of varying ages. This provides fish for planting every month of the year. If fish do not move from pond to pond, it is possible to have screens only at every third or fourth pond.

Waste material from the fish accumulates in the ponds. As the fish grow larger and become more active the waste material drifts down to the lower end of the concrete raceway type rearing ponds and accumulates against the checkboards. A piece of corrugated fiberglass mounted on two vertical pieces of 2 x 4 placed against the upstream side of the checkboards so the fiberglass will be 3½ inches above the pond bottom and extend above the water surface will force all water to pass under the fiberglass and up between it and the checkboards. This current will move the accumulated material from pond to pond which will virtually make the ponds self-cleaning of all solid waste material that reaches the lower end of the pond (Figure 49). The waste material will not move down in an earthen pond as well as it will in a concrete pond.

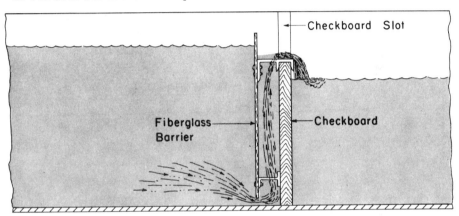

SIDE   SECTION   VIEW

FIGURE 49—Illustration of partial barrier placed on upstream side of checkboards, which prevents accumulation of waste material at lower end of raceway pond.

Pond Screens

The pond screen in common use is the perforated plate aluminium screen (Figure 50). Some of these are mounted on 1¼ inch OD x .125 inch T aluminium tube frame with ³⁄₁₆ inch blind rivets at 4 inch centers. Others are mounted on redwood frames. Some hatchery men feel it is easier to make the redwood frames fish tight than the metal frames due to irregularities in the concrete. Aluminium plates with many sizes of holes and oblong slots are used varying from ¹⁄₁₆ inch holes to a ½ inch x 1½ inch staggered slots and a rack with vertical bars used with brood fish with 1 inch spacings between the bars. California has more or less standardized on the following specifications for the average hatchery rearing ponds:

.063 aluminium plate—(20 gauge)
⁵⁄₆₄ inch round holes—½ inch staggered .... 36% open space
⅛ inch x ¾ inch slots—side staggered........................ 41% open space

.032 aluminium plate—(14 gauge)
³⁄₁₆ inch x 1½ inch slots—side staggered .... 42% open space
⅜ inch x 1½ inch slots—side staggered.................... 46% open space

FIGURE 50—Aluminum plate-type pond screens. **A**, round holes; **B**, staggered slots; and **C**, non staggered slots. *Photographs by George Bruley, 1973.*

These perforated aluminium plate screens are available from a number of venders. A few of these in California are:

California Perforating Screen Co.
681 Market Street
San Francisco

Diamond Metal Sales
17915 So. Figueroa Street
Gardena

Aluminium Service Co.
1401 Middle Harbor Road
Oakland

If there are many leaves or other debris accumulating in the water, rotary screens are desirable particularly in upper and lower ends of the pond series. Two types of rotary screens are used in California. The screen drum (12 to 16 inches in diameter) usually is covered with galvanized 8 or 4 mesh hardware cloth and turned by a paddle wheel inside the drum. The drum turns in the opposite direction and carries the debris over the screen where it is washed off on the downstream side (Figure 51).

FIGURE 51—Water driven rotary screen. The paddle wheel turns the opposite direction of the screen drum. The frame on each end of the screen fits in the keyways in the pond walls. A similar frame member across the bottom makes the screen fish tight in the pond. *Photograph by Jerry Eskew, 1973.*

A

B

FIGURE 52—Electrically driven rotary screens. **A**, screens in operation with electric motor on top of the frame; **B**, details of frame, chain drive and motor on screen. *Photographs by George Bruley, 1973.*

The other type rotary screen is used in a deeper pond with a drum from 2 to 3 feet in diameter. It is turned by an electric motor through a chain drive and passes debris over the top to wash off on the other side (Figure 52).

Mechanical Crowder on Concrete Ponds

A mechanical crowder has been constructed (Figure 53) which runs on top of the concrete pond walls and is propelled by a gasoline engine. The crowder has an adjustable Wilco Grader which can be mounted on the crowder frame and adjusted to any size opening to crowd fish within a pond for loading, moving to another pond, or for grading.

FIGURE 53—A mechanical crowder used in concrete rearing ponds with an adjustable Wilco grader mounted on the crowder frame. Note the wiper blades on each side of the frame to clean algae from the walls. *Photograph by George Bruley, 1973.*

The crowder frame fits very snugly between the pond walls and on the bottom with wiper blades that clean the algae from the walls and bottom as the crowder proceeds down the pond series. The fish swim through the frame and under the crowder, remaining in the same pond. The frame is raised and lowered hydraulically to pass over screens and/or checkboards in the pond series. Ponds can also be cleaned very readily with this machine. The crowder is moved from one pond series to another by running it onto a small car on rails at the end of the pond series and pushing the car to a point where the crowder can be put on any series desired (Figure 54). Lighter weight models with aluminum frames can be moved by forklift.

Advantages of Concrete Ponds

The many advantages of the concrete raceway type rearing pond over earthen ponds is summarized: (1) no weed growth or bank erosion, (2) no structural maintenance, (3) swimup fry can be put directly into ponds from incubators without excessive loss and the need for hatchery troughs is eliminated, (4) good circulation, (5) good control for chemical flush treatment of disease, (6) use of mechanical crowder for cleaning ponds, grading, and herding fish together for loading tank trucks or pumping into other ponds through a pipe, (7) good accessibility for tank trucks, fish pump, and mechanical pellet blower, (8) easily

FIGURE 54—A small car runs on steel rails at the head of the pond series so the crowder can be moved from one series to another. The crowder is run onto the car which is moved to the series desired and then run onto the pond series. *Photograph by George Bruley, 1973.*

adaptable for use of automatic feeder, (9) compact unit in small area to simplify working conditions, (10) self-cleaning devices aid in moving solid waste material from the ponds, (11) smaller area simplifies bird control problem, (12) seines are not required for catching and moving fish, and (13) simplifies overall management.

This all adds up to mechanizing many of the hatchery operations with a great saving in labor which reduces the unit cost of producing fish.

FIGURE 55—Wires strung over the pond series spaced about 2 feet apart have discouraged sea gulls from flying into the ponds. (Wires can also be seen in Figure 47). *Photograph by George Bruley, 1973.*

Spawning Channels

Spawning channels are constructed in conjunction with hatcheries or below impassable dams to provide natural spawning areas for salmon (Figure 56). The parent fish generally enter the channel through a fish ladder and a counting rack and are counted so the number of fish does not exceed the capacity of the spawning area.

The channels can be constructed from a few hundred feet to 1 or 2 miles in length as needed and the location will permit. They can cover a large area or be in a compact unit of parallel channels with the water running in opposite directions. The width is generally from 20 to 50 feet depending on the amount of water available to meet the water depth and velocity criteria. The criteria for spawning

FIGURE 56—Two spawning channels which provide natural spawning areas for salmon. *Photograph by George Bruley, 1973.*

channels varies with the species of fish, location, and general conditions. The following criteria are considered as generally suitable for king salmon:

Slope of bottom	—	1 to 2 feet per 1000 feet
Gravel size	—	¾ to 6 inches
Gravel depth	—	18 to 30 inches
Water depth	—	12 to 18 inches
Water velocity	—	1½ to 2½ feet per second measured at 0.3 feet above gravel surface
Resting pools	—	10 to 20 feet long and 5 feet deep, spaced about 250 feet apart for green fish to rest until they ripen.

The area required for each king salmon female to spawn in a channel varies with the location and general conditions but is usually figured at approximately

60 square feet or more of spawning gravel. This is the basis to determine the capacity of a specific spawning channel.

The young fish can migrate back into the parent stream enroute to the ocean or can be trapped at the lower end of the channel and measured and counted to evaluate the program and trucked to another release site if desired.

AQUATIC VEGETATION IN PONDS

Aquatic vegetation can usually be divided into three general groups: algae (suspended and mossy filamentous), emergent vegetation (tules and cattails), and submergent vegetation (plants rooted to the bottom that do not extend above the water surface). All are a hindrance to fish removal, feeding, access, and general pond management. In raceway ponds with a fair current, aquatic vegetation other than algae normally does not become too bothersome. This is particularly true with the concrete lined ponds. Raceway ponds do not lend themselves very well to weed eradication programs. In the deep, still-type pond, aquatic vegetation usually grows rather profusely, but control measures are more easily effected. Under certain conditions, plants regulate the amount of oxygen available to the fish in a pond and govern the number of fish which can be carried.

Photosynthesis

Photosynthesis is the process by which plants manufacture food. Plants that are able to carry on photosynthesis contain a green substance, chlorophyll. The sun furnishes the energy required for plants to produce certain chemical combinations. In the process, oxygen is freed and escapes from the tissues of the plant by diffusion. Thus, aquatic vegetation in fish ponds helps build up the oxygen content in the water during periods of sunlight and at such times may even be beneficial. The process, however, is reversible, and during periods of darkness or overcast days, when sunlight is absent, oxygen is taken from the water by plant growth. Oxygen depletion, due to overabundant plant growth, with resultant fish kill, it not uncommon.

Aquatic Vegetation Control

The task of controlling aquatic vegetation has been made considerably easier during the past few years by the development of new chemicals known as herbicides. The ideal herbicide should have the following qualifications: (1) it must destroy or depress aquatic vegetation at concentrations not lethal to fish or other desirable aquatic organisms, (2) it must not adversely alter the water or substrate as a desirable biological environment, (3) it must be dependable under a wide variety of physical and chemical conditions, (4) it must be safe to handle and apply, and (5) it must be economical to use.

Although a number of new and promising chemicals have appeared on the market in recent years, the ideal aquatic herbicide, as described above, has not been developed.

Most of the chemicals are poisonous to both fish and/or other animals when used above the recommended dosages.

The use of many chemicals that were used a few years ago is now restricted. In most cases a permit must be obtained from the County Agricultural Commissioner before chemicals can be put into the water. The Commissioner's office should always be contacted.

Chemicals should not be used without expert advice regarding concentrations and their effect upon man and wildlife. Information on aquatic weed control may be obtained from the County Agricultural Commissioner or the Agriculture Extension Service of the University of California at Davis.

Details on the use of most chemicals may be secured by writing the manufacturers. Even though effective chemicals are available, control of aquatic vegetation is quite complicated. Concentrations must be kept harmless to fish life, yet be strong enough to kill vegetation. Before attempting chemical control, one must determine the strength of solution usable without endangering the fish. Normally, warmwater fish can withstand higher concentrations than trout and salmon.

To apply the correct amount of chemical, after the concentration has been determined, it is necessary to determine the volume of water to be treated. The following formula should be used to compute the volume and amount of chemical required.

1. Determine volume of pond or reservoir in acre-feet:

$$\frac{\text{Length in feet} \times \text{width in feet}}{43,500} \times \text{average depth} = \text{number acre-feet}$$

2. Multiply number of acre-feet by 0.3 to convert to millions of gallons.

3. Determine the amount of chemical to be used in pounds per million gallons.

4. Multiply pond volume (expressed in number of million gallons or fraction thereof) by pounds of chemical required for proper species.

Example: Pond 100 feet × 200 feet × 10 feet *average* depth contains trout.

a. $\dfrac{100 \times 200}{43,500} \times 10 = $ approximately 5 acre-feet.

b. 5 acre-feet × 0.3 = 1.5 million gallons.

c. 10 pounds of chemical used per million gallons is tolerance for trout.

d. 10 × 1.5 = 15.0 pounds of chemical required for pond treatment.

Normal tolerance to various herbicides for trout and warmwater fish is as follows:

	Concentration in p.p.m.	
	Salmonids	*Centrarchids*
Copper sulfate	0.1	1.0
Delrad 50 and 70	0.5	1.0
Karmex W	1.0	5.0
Dalapon	*	*
Aminotriazole	*	*
Sodium Arsenite	4.0	7.0
Silvex	*	*

* Not fully determined.

The strength of solution to be used can best be determined by contacting the company supplying the chemical.

The effectiveness of most herbicides is highly variable and is dependent on temperature and alkalinity of the water and density of the aquatic growths. Toxicity to fish is likewise affected. Before large-scale control measures are put into effect, it is well to experiment on a small scale with the chemical to be used and determine the concentration which will most satisfactorily do the job at the particular location.

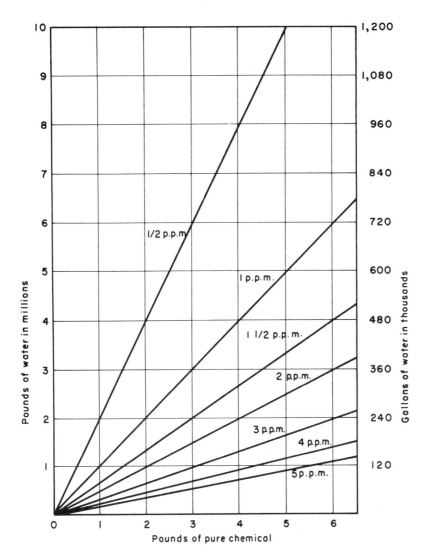

FIGURE 57—Concentration chart.

Treat plants (except tules and cattails) preferably while they are young, tender, and growing rapidly. Do not wait until they become a nuisance, since then it may be too late for effective control and a great deal of time and expense may produce no more than moderate success. There is also the danger of oxygen depletion, caused by decaying vegetation, when luxuriant plant growth is killed. Treatment of tules and cattails should take place when plants are mature, since success of spraying is proportional to the leaf area contacted by the chemical.

In using herbicides, the user is cautioned against the danger of destroying valuable plants nearby through air drift. This is particularly true in the case of chemicals which are applied by surface spraying.

TROUT AND SALMON NUTRITION

Almost no phase of fish culture is more important than nutrition and feeding. For many years, it was quite generally assumed that beef liver was a complete diet for trout, and beef liver was fed in trout hatcheries in greater amounts than any other food. Fortunately, rather large amounts of condemned livers (livers condemned because of fluke infestation), which kept the price within a reasonable range, were available to hatcheries situated west of the Rocky Mountains.

As larger trout were produced, it became evident that beef liver, while it was one of the best trout feeds available, was not a complete trout food and it became common practice to supplement a liver diet with other packing-house products, such as lungs, hearts, and spleens. Early fish culturists in this country suspended freshly killed beef heads over their ponds. After a few days maggots would fall from the heads into the ponds. This is probably one of the earliest attempts to provide a natural food for hatchery fish. Such products as ground whole fish, both ocean and freshwater varieties, fish cannery by-products, and numerous cereals were also used to supplement hatchery diets of packinghouse products. A balanced ration of dried meals, vitamins, and supplements in pelletized form of various sizes is now the conventional diet in hatcheries throughout the nation.

The nutritional requirements of trout and salmon are not yet fully determined. Fish nutrition, when compared to that of other domestic animals, is still in its infancy. Obvious reasons for this are the lack of work in this field and the difficulty of working with fish. The amount of research in trout nutrition is constantly increasing. In future years the trout production industry may be abreast of other fields in nutritional knowledge. In the meantime, many of the requirements of trout are based on the dietary needs of other animals. In general, this is a safe assumption. However, fish are cold-blooded animals, their body being of the same temperature as the water surrounding them, and their ability to assimilate food under varying conditions must be taken into account.

The intestinal tract of a trout is that of a typical carnivore. Both small and large intestines are very short, and the total length of the digestive system is not great enough to allow any important synthesis of vitamins by bacteria in the intestines. Most of these requirements must be supplied in a digestible form in the food.

Realization of these shortcomings in a trout's digestive system is a long stride forward in understanding its food requirements.

Present nutritional knowledge has been built around five principal groups of nutrients. They are fats, carbohydrates, proteins, minerals, and vitamins.

Fats

The three sources of fat used to meet bodily requirements are fat of the diet, fat produced by excess dietary protein, and fat produced by excess carbohydrates.

Most dietary fat is changed to fatty acids and glycerol in the small intestine before absorption. The ease of this process is primarily dependent on the melting point of the fat. Soft fats are easier to digest than very hard fats. Digestibility varies from 70 to 90%, depending on the melting point and body temperature.

Hard fats may retard or prevent the digestion of protein and carbohydrates by coating their molecules. This insulates them from the action of the protein- and carbohydrate-digesting enzymes and acids. Small trout have special difficulty in digesting hard fat. It is believed that hard fats also lessen a trout's ability to adjust to water temperature changes.

The body uses fats for energy, insulation, and cushioning of vital organs, and as an internal lubricant. They are also stored for future use.

Fats aid in the absorption of the fat soluble vitamins, which are necessary for normal health and growth. Phospholipides are kinds of fat which contain phosphorus, fatty acids, and choline, or other important bases.

There are two types of fat deposits in the body: one is the natural fat of the animal manufactured from the protein and carbohydrate of the diet; the other is deposited from fat in the diet. Fat deposited from dietary fat is similar to that of the original food. Excessive fat deposits may be caused either by overfeeding or by high fat content of the food.

Excessive dietary fat can cause body damage resulting in death. Important food conversions take place in the liver; fatty infiltration of the liver may cause anemia resulting in the death of the fish. Kidneys damaged by excessive fat deposits may result in edema, i.e., an accumulation of water in the body.

The principal sources of fats are fish meal and oil, cottonseed meal, rice bran, fresh fish and meat, and meat and bone scrap.

Cod liver oil has been found to increase growth under certain conditions, but it is not clear whether this effect is caused by the lipides themselves or the vitamin A and D present in the oil. Due to possible kidney and liver damage, caution should be used in feeding cod liver oil in cold water. Oxidation of fish oils such as cod liver oil, when mixed into the diet, will cause destruction of vitamin E.

Insofar as is now known, the trout diet should contain not less than 5% and not more than 8% fat.

Minerals

Minerals are generally considered important in building strong bones. The important part they play in the functional activities of the body is not widely appreciated. Blood circulation, respiration, digestion, and food assimilation, as well as excretion, are dependent on the presence of minerals in suitable compounds. The major minerals important in trout and salmon nutrition are calcium and phosphorus (Table 8).

TABLE 8
Important Nutritional Minerals

Major	Use	Trace	Use
Calcium	Bones, teeth, blood coagulant	Cobalt	Red blood cells
Phosphorus	Bones, teeth	Iron	Red blood cells
		Copper	Red blood cells; enhances enzyme action
		Magnesium	Bones, teeth
		Sodium	Osmotic cell pressure
		Chlorine	Osmotic cell pressure, digestion
		Potassium	Osmotic cell pressure
		Manganese	Growth
		Fluorine	Teeth, bones
		Iodine	Regulates metabolism

Usually minerals are needed in only small amounts. While calcium, phosphorus, and iron are used in building the body and blood, most of the minerals function as catalysts.

Trout have the ability to absorb calcium, cobalt, and phosphorus from the water. Enough for bodily needs may be absorbed if there is enough present in the proper form. The amount absorbed from the water varies in proportion to the amount in the water.

Calcium and phosphorus in the ratio of 2 to 1, respectively, are the major minerals used in forming the bones and teeth. Fluorine and magnesium are trace minerals involved in the structure of the body.

Iron, cobalt, and copper are trace minerals used to build red blood cells. Iron and cobalt are combined in the red blood cells. Copper acts as a catalyst to aid in the assimilation of iron. A deficiency of any one will cause anemia.

Sodium, chlorine, and potassium regulate the osmotic pressure of the body cells. Body fluids contain about 90% of the body content of these minerals.

Minerals involved in the special body processes act as catalysts and, with the exception of calcium and phosphorus, are considered to be trace minerals. Chlorine aids digestion, copper increases enzymatic action, iodine helps regulate metabolism, and calcium is a blood-coagulating agent.

To illustrate the limited knowledge of trout nutrition, only iodine has been proven to be essential and only salt has been found to be harmful when fed in excessive amounts.

Fish, bone, kelp, and meat meals; dried skim milk; fresh meats; and fish are good mineral sources.

Vitamins

The understanding of vitamins has increased rapidly in the last few years. There are now 16 generally recognized vitamins (Table 9). Probably there are more to be discovered.

TABLE 9

Vitamins

Vitamins	Minimum daily requirement in mg. per kg. body weight	Deficiency symptoms
Fat soluble		
A	Unknown	Unknown. Believed to cause cataracts, retarded growth.
D	Unknown	Unknown.
E	Unknown	Unknown.
K	Unknown	Unknown.
Water soluble		
B¹ thiamin	0.150	Mortality from shock and fright; poor appetite; instability; convulsions; pale livers.
B² riboflavin	0.44-0.68	Blindness; hemorrhagic eyes, nose, and operculum.
Pantothenic acid	0.97-1.25	Western gill disease.
Pyrodoxine	0.225-0.250	Nervous disorders; light spots on the liver.
Inositol	Unknown	Poor growth; distended stomach; degenerated fins.
Biotin	0.0433-0.00678	Anemia; poor growth; blue slime.
Folic acid	0.00292	Anemia; poor growth.
Niacin	3.0-4.0	Swollen, unclubbed gills; poor growth; back peel.
Ascorbic acid-C	Unknown	Hemorrhagic liver, kidney, and intestine (suspected).
Vitamin B¹²	Unknown	Unknown. Probably not required.
Para-aminobenzoic acid	Unknown	Unknown. Probably not required.
Choline	Unknown	Hemorrhagic liver, kidney, and intestines; fatty infiltration of the liver.

Some vitamins are found in more than one form, notably A and D. Each vitamin performs functions not wholly duplicated by any other, yet the actions are interrelated.

It is known that only minute amounts are needed and that they act as catalysts, making it possible for the body to utilize the other diet components.

Vitamins are classified as fat-soluble and water-soluble. In general, fat-soluble vitamins can be stored in the body and water-soluble vitamins cannot.

Of the 16 known vitamins, 10 are known to be essential to trout. However, until further work is done, it should be assumed that all vitamins are essential.

It is generally conceded that unknown diseases and unusual severity of bacterial diseases may be caused by vitamin deficiences. Other than their effect on the general condition of the fish, they are not believed to have any effect on virus diseases.

To add to the hatchery managers' problems, only four of the ten essential vitamins have definite external deficiency symptoms. Symptoms of the others are shown by retarded growth rate and internal damage. There is considerable overlapping of symptoms.

Proteins

The main component of the body organs, soft tissues, and body fluids is protein. Blood meal, for instance, contains 85% protein. Beef liver meal contains 66% protein.

Proteins are made up of amino acids, which consist of carbon, hydrogen, oxygen, and nitrogen. The nitrogen level is fairly constant at 16%. The amino acids are present in widely varying amounts in different proteins. Eighteen amino acids have been identified. Ten of these are probably essential in trout food (Table 10). The other eight may be manufactured in the body, or are not required.

TABLE 10
Known Amino Acids

Essential	Other
1. Arginine	1. Tyrosine
2. Histidine	2. Glycine
3. Isoleucine	3. Alanine
4. Leucine	4. Aspartic acid
5. Lysine	5. Cystine
6. Methionine	6. Glutamic acid
7. Phenylalanine	7. Proline
8. Threonine	8. Serine
9. Tryptophan	
10. Valine	

The term biological value of a protein indicates the degree of digestion and amino acid balance of a protein. The more nearly a protein resembles the protein found in the body of the animal being fed, the more it is utilized.

Digestion splits proteins into amino acids. The acids then pass through the intestinal wall into the blood stream. They are then carried to the liver and to other body cells, where they are reformed into proteins of various kinds.

Proteins are used primarily for growth. Excess protein may be used for energy or deposited as fat. In most animals, fats and carbohydrates are used for energy. However, if the diet is low in these, protein will be used to supply energy.

No amino acids are stored in the body. They must be supplied in the proper balance each day and they should all be fed at the same time.

While not fully understood, the amino acid content of proteins is known to be damaged by high temperature. It is also known that protein content may be lost by leaching.

The desirable level of protein in the trout diet is believed to be 28%. There is some evidence that a diet containing 50% protein may be fed to advantage.

Protein deficiency symptoms are even more indefinite than vitamin deficiency symptoms. However, they occur much more quickly. The only protein deficiency symptoms which have been described are lack of appetite, reduced activity and growth, and fish staying near the top of the water.

Animal concentrates, fresh meat, and fish are excellent sources of protein. Plant concentrates, especially seeds and grains, are also good sources of protein, although their use is complicated by their high carbohydrate content and their amino acid patterns.

Carbohydrates

Carbohydrates are made up of hydrogen, carbon, and oxygen. Unlike proteins, they do not contain nitrogen. They are used as energy, temporarily stored as glycogen (animal starch), or formed into fat.

Compound carbohydrates are digested into simple sugars before they are absorbed. Their availability is dependent on a fish's ability to digest them.

Carbohydrates in the body are found in the form of glucose (sugar) and glycogen. Glucose is deposited in the fluids and cells of the body, and glycogen in the liver and muscle tissues.

Excess carbohydrates in the diet will cause swollen bodies and deposit excess glycogen in the liver, resulting in swollen, light colored livers. Mortality will be high.

Not over 9 to 12% of digestible carbohydrate should be included in a diet for salmonids. Practically all trout and salmon diets containing dry food exceed this percentage of raw carbohydrates. This is possible because all of the carbohydrates are not digested. Often only a small part is available to the fish. For example, only a minute fraction of the carbohydrate content of distillers' solubles is digested.

In general, the more complex the carbohydrate, the harder it is to digest. However, cooking makes carbohydrates more digestible. A satisfactory diet may be ruined by cooking simply because the carbohydrates become more digestible.

Major sources of carbohydrates are plant products. Meats contain only minor amounts of carbohydrates.

Interrelationships

Up to this point, nutrition does not sound too complicated; one feeds a diet containing the right amount of fats, minerals, carbohydrates, and vitamins; if a deficiency appears, the lacking component is added and the deficiency is corrected.

To some extent, this is true. However, a great many interrelationships between vitamins, between minerals, and between amino acids, as well as between members of these groups and of other groups, such as proteins, fats, and carbohydrates, have been discovered. Sometimes the interrelationship hinges on the intervention of a third factor. Some interrelationships affect the synthesis of vitamins, others the preservation, destruction, or functioning of vitamins.

One example is the relationship between vitamin D, calcium, and phosphorus. All are needed to build strong bones. Vitamin D's function is to prevent rickets. However, if there is not enough calcium or phosphorus in the diet or if they are not in the proper proportion, vitamin D cannot prevent rickets. An excess of calcium will increase the demand for vitamin D until a level is reached at which it is impossible to supply enough to prevent rickets.

Some materials act as inhibitors. Linseed oil meal may cause a pyrodoxine deficiency even though the diet contains a high level of pyrodoxine.

A shortage of any essential amino acid will limit the effectiveness of the others present to the level of the lowest one. The same is believed to be true of vitamins. Excess of some diet factors may cause an imbalance which will cause the whole diet to fail.

The synthesis of the B group of vitamins and amino acids in trout is probably negligible. Synthesis of these groups depends on intestinal bacteria, which are not believed to be important in the intestines of trout.

All of this leads to the conclusion that it is not so much what the diet contains that is important, but what it accomplishes in the body. Nutrients are rarely completely digested and frequently the digested fraction of the nutrient is not completely utilized.

FOOD STORAGE AND PREPARATION

The best foods available may lose their value through improper storage and preparation. Millions of dollars worth of synthetic vitamins, purchased annually

FIGURE 58—Feed room at Mt. Shasta Hatchery, 1914. Diet consisted of ground beef liver, clabbered milk, and cooked wheat middlings. Bags in foreground contain wheat middlings.

by consumers, are lost before feeding. Microorganisms, enzymatic action, and oxidation, encouraged by unfavorable conditions, contribute to the destruction of the nutrients contained in foods.

The action of bacteria, yeasts, and molds causes spoilage and deterioration. These microorganisms utilize the soluble constituents of the food or secrete enzymes which cause decomposition.

Enzymatic action results in a process known as autolysis. Autolysis resembles digestion, except that digestion is carried out by enzymes in the tissues of living material. In dead tissue the action continues and results in tissue disintegration.

Proteins, fats, and vitamins are the nutrients most subject to destruction in storage and preparation. These are also the most important items in a diet for salmonids.

Proteins are subject to destruction by microorganisms, autolysis, and heat. Microorganisms and autolysis cause decomposition, leading to alteration or destruction of the protein content of the food. High temperature alters the protein content, reducing the biological value of the protein.

Fats are altered or destroyed by autolysis and oxidation. Autolysis increases the acidity of fats. Rancidity, resulting from oxidation, can make fats indigestible and may produce fatal toxins. Vitamins A, C, and E are destroyed by oxidation. In addition to destruction by oxygen, microorganisms, and autolysis, vitamins may be unfavorably affected by light, heat, and moisture.

Freezing and drying are the principal methods employed for the preservation of fish foods. Refrigeration is the most satisfactory method of preserving raw animal products. Autolysis and the action of microorganisms are inhibited by cold temperature. The degree of coldness has a direct bearing on the speed of these processes.

Meats and fish should be quick-frozen at $-35°$ F and stored at $-10°$ F. At this temperature, autolysis is minimized and oxidation is greatly reduced. Deterioration does continue, however, and considerable value may be lost after 90 days in storage.

Retention time before freezing and the effects of different storage temperatures affect the quality of fish (Figure 59).

The upper curves show a 15% quality loss due to freezing when frozen immediately after taking. The rate of deterioration to zero quality is shown as 12 months at $-20°$ F temperature, increasing to 6 months at $0°$ F storage temperature.

The middle curves indicate the quality relationship when fish are held on ice for one week prior to being frozen. These will enter the freezer with a quality score of 66⅔%, leave in the frozen state at 50⅔%, and exhibit a considerably shortened storage life. The lower four curves indicate the doubtful practicability of freezing fish that have been held 2 weeks on ice.

The manner in which animal products are handled prior to freezing is of great importance, if the maximum value is to be preserved. The protein quality and vitamin content of meat products which have been allowed to decompose partially before freezing are reduced. Fish products in which decomposition has occurred before freezing may prove to be toxic.

Thawing and refreezing results in reduction of the vitamin content, due to rupture of cell walls. Slow thawing is responsible for a high loss of water-soluble vitamins. Prolonged retention of any ground product after thawing results in

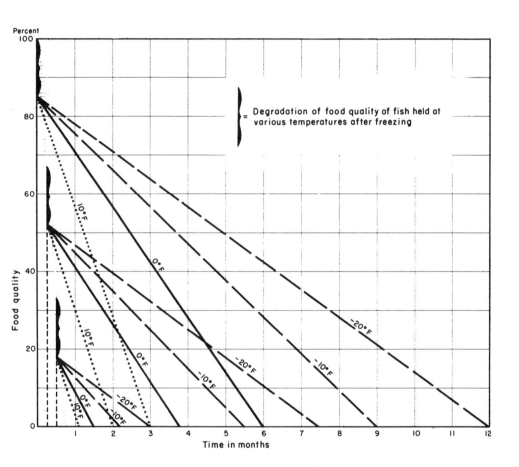

FIGURE 59—Degradation of food quality of fish held at various temperatures after freezing.

reduced food value. No more food should be ground or thawed at the hatchery than can be used in 48 hours. To prevent excessive loss of diet value, ground meat and fish products should be stored after thawing at as cold a temperature as possible without fusion of the particles by freezing. Usually, 30° F is satisfactory.

The conversion of raw products to meals by the application of heat is used extensively. Dehydration by heat stops autolysis, but accelerates oxidation. The proteins may be altered by high temperatures. Some of the vitamins are destroyed by heat. Precooking before dehydration causes a loss of the water-soluble vitamins. Low temperature dehydration processes have been developed. It is desirable to use meals processed by this method in dry food formulas.

Once a product has been reduced to meal, the vitamin content is relatively stable. However, when various meals are mixed together, some of this stability is believed to be lost. Dry food should be stored in a cool, dry place until used. It should not be stored longer than 60 days.

The dried meals prepared in pelletized form are stored in feed bins (Figure 60). The vendor delivers the food in bulk quantities and loads the feed bins at the hatcheries. Each hatchery has two or more bins so various sized pellets can be stored.

FIGURE 60—Feed bins, for storing bulk delivery of various sizes of dry fish food. *Photograph by George Bruley, 1973.*

The mobile pellet feeders described later are loaded by gravity flow from the feed bins. Small grades of dry food are received at the hatcheries in sacks and stored in a cool, dry place. These grades of food are used only in small quantities for the smaller fish.

The Oregon Moist feed is also received in sacks and stored at temperatures of 0° F. This feed is used for salmon and steelhead and for starting trout to feed for about 10 days or 2 weeks.

HATCHERY DIET INGREDIENTS

The ingredients of a trout diet must be available in sufficient quantities to supply the demand and produce good growth with a minimum mortality. No single component is as satisfactory as a combination because no single component contains all the nutritional requirements of trout and salmon (Table 11).

TABLE 11
Analysis of Some Hatchery Foods

Food	Percent protein	Percent fat	Percent carbo-hydrate	Percent water	Percent ash	Percent fiber
Fresh meats						
Heart						
Beef	14.8	24.7	0.9	53.2	0.9	----
Sheep	16.9	12.6	----	69.5	0.9	----
Pig	14.4	2.9	----	79.5	----	----
Kidney						
Beef	13.7	1.9	0.4	74.0	1.0	----
Sheep	15.3	4.1	----	77.9	----	----
Pig	15.5	4.8	0.7	75.0	1.2	----
Liver						
Beef	20.2	3.1	2.5	72.3	1.3	----
Sheep	23.1	9.0	5.0	61.2	1.7	----
Pig	21.3	4.5	1.4	72.8	1.4	----
Horse	20.0	3.0	2.5	----	1.3	----
Lung						
Beef	16.4	2.8	----	79.7	1.0	----
Sheep	20.2	2.8	----	75.9	1.2	----
Muscle						
Horse	19.3	5.5	----	73.6	----	----
Beef	20.0	7.0	----	65.3	----	----
Tripe						
Beef	11.7	1.2	----	86.5	0.3	0.3
Spleen						
Beef	18.0	2.3	----	75.2	1.4	----
Pig	17.0	1.9	----	78.0	1.4	----
Viscera						
Horse	19.8	1.2	----	77.0	1.1	----
Fresh fish						
Alewives	10.0	2.4	----	----	0.8	----
Carp	10.0	15.0	----	----	0.5	----
Herring (sea)	11.2	3.9	----	----	1.6	----
Mackerel	10.0	4.2	----	----	0.7	----
Salmon	15.3	8.9	----	----	----	----
Shad	9.4	4.8	----	----	0.7	----
Smelt (sea)	18.7	1.3	----	----	----	----
Whiting	18.8	4.0	----	71.0	5.4	1.7
Canned fish						
Mackerel	19.6	8.7	----	----	1.3	----
Salmon	19.5	7.5	----	----	----	----
Shrimp	25.4	1.0	0.2	----	2.6	----
Animal meal						
Blood	82.2	1.2	0.7	----	3.3	0.6
Buttermilk	33.8	5.6	41.9	7.8	10.5	0.3
Bone	7.1	3.3	3.9	----	81.3	0.8
Liver	67.2	14.6	2.4	----	7.5	1.9
Meat	55.0	12.0	1.2	----	25.0	2.2
Meat crisp	75.0	----	----	9.7	----	----
Skim milk	34.8	0.9	50.1	6.2	8.0	----
Whey	12.5	0.7	72.1	----	9.7	----
Fish meal						
Crab	36.5	2.9	7.4	----	39.5	5.7
Dog fish	76.3	8.4	----	11.0	11.1	----
Herring	72.6	----	----	----	----	----
Menhaden	61.0	7.0	0.8	----	21.0	2.4
Salmon	55.4	----	----	----	----	----
Salmon egg	39.3	8.7	----	9.3	10.7	----
Salmon viscera (pink)	59.6	17.0	----	5.7	11.3	----
Salmon viscera (red)	60.1	24.3	----	5.7	----	----
Shrimp	42.1	2.2	1.4	----	33.8	9.3
Sardine	64.5	9.8	3.8	----	15.2	0.2
White fish	54.5	8.8	4.6	10.5	22.0	1.0
Plant meal						
Apple pumace	4.5	5.0	62.1	----	2.2	15.6
Beet pulp	9.0	0.8	59.9	----	3.5	18.8
Bran	16.0	4.4	63.2	----	6.3	9.5
Corn gluten	26.4	2.5	48.4	----	6.1	7.1
Cottonseed	38.0	8.0	39.9	----	6.4	10.5
Kelp	5.6	0.7	43.7	----	33.2	7.5
Oat flour	15.0	5.8	66.7	----	1.9	2.6
Oatmeal	16.0	6.5	67.6	----	2.0	2.8
Peanut	44.8	10.2	33.6	----	4.8	18.9
Red dog flour	16.8	4.1	65.5	----	2.5	2.4
Rice bran	12.8	13.4	41.1	----	10.8	13.0
Soy bean (expeller)	49.9	6.2	26.4	----	5.5	5.1
Soy bean (solvent)	46.4	1.6	31.7	----	6.0	5.9
Wheat middlings	17.8	5.0	62.8	----	3.7	4.4
Wheat shorts	17.4	4.9	62.8	----	4.4	6.2

Dry meals are used extensively in both trout and salmon rations. They have been used as a supplement mixed with meats and more recently as a complete ration in pelletized form. In some cases, the use of dry meals has resulted in excessive losses.

Vegetable products are considered inferior to animal products as sources of protein. They contain relatively high levels of more-or-less digestible carbohydrates and must be fed with caution. Cereal meals are high in B complex vitamins and are good sources of protein and minerals.

Dried skim milk, liver meal, and crab meal are high in vitamin content.

The method by which a meal is prepared, with special emphasis on temperature, affects its efficiency. Meals dried at temperatures below 145° F produce better results than flame-dried meals. High temperatures reduce the biological value of the protein by alteration. High temperatures also reduce the vitamin content.

Complete dry food trout rations are a mixture of meat, fish, and vegetable meals, usually with minerals and purified vitamins added. Great care must be exercised in balancing the amino acid and vitamin content. Properly balanced, the dry food diet will produce healthy, fast-growing fish. Not only is it necessary to have the right combination of meals, but the quality of the meals must be carefully controlled. Only tested brands of commercially prepared dry foods are used in California hatcheries. This, in effect, shifts the responsibility for quality control to the manufacturer. No dry food is purchased until it has undergone performance tests.

Experiments are carried on at various hatcheries on a continuing basis to test the quality of dry foods produced by manufacturers to determine whether they meet performance requirements.

Some of the more common ingredients of complete dry food diets are dried skim milk; herring, whitefish, blood, liver, soybean, and cottonseed meals; wheat middlings; and other vegetable products.

Spray-processed dried skim milk contains 37% protein, 1% fat, and 49% carbohydrate. All three are in a readily absorbent form for trout and salmon. The mineral and vitamin content is extremely high. The high digestible carbohydrate content limits the amount which can be used.

Whitefish meal from the offal of cod, haddock, and other whitemeated fish contains about 68% protein and 2% fat. The mineral content and the biological value of the protein are high.

Other species of fish meal have also been used. Fish that have a high oil content do not produce good meals because of the tendency of the fat to turn rancid in storage.

Blood meal and meat meal have a high protein content of low biological value. They are used to supplement the protein value of other ingredients.

Vegetable meals are usually not mechanically dried, but are either ground or cooked. Although their protein quality does not equal that of animal meals, they are included to cheapen the diet and supplement the amino acid content of the other ingredients. Usually, they are high in vitamin B content.

Both cottonseed and soybean meals have relatively high protein content, 38% for cottonseed and 44% for soybean. Cottonseed meal is higher in fat and carbohydrates than soybean meal.

Wheat middlings contain about 17% protein and 63% carbohydrate, but are rich in B complex vitamins. The carbohydrates are of complex structure and are relatively indigestible by trout and salmon.

Other vegetable products are used as mineral and vitamin supplements. For example, kelp meal is particularly high in iodine content.

The choice of foods supplied to California trout hatcheries is limited by availability and price. The initial price of each product is important; however, the price per pound of fish produced on the food is usually the deciding factor.

Oregon Moist feed produces better results in feeding salmon and steelhead than the conventional dry food pellets. It is used extensively in the salmon and steelhead hatcheries in California. In some locations, the feed can be changed to the conventional dry food after a few months with equal results, particularly with steelhead.

Trout start feeding more readily on Oregon Moist feed than on the conventional dry food; therefore, it is common practice to feed Oregon Moist for the first 10 days or 2 weeks to start trout to feed well, then the diet is changed to regular dry food.

The oil content in the Oregon Moist feed keeps it soft. It is necessary to keep this food frozen in storage, preferably at 0°F .

The Oregon Moist feed is prepared to meet specifications prepared by the State of Oregon. The food is produced as a mash for use as a starter feed for fry beginning to feed, and in pelletized form for older fish.

The diets contain three basic components: (1) meal mix, (2) vitamin premix, and (3) wet mix. The meal and wet mixes vary between mash and pellets, but the vitamin mix is the same for both diets.

The meal mix for Oregon Mash consists of herring meal, wheat germ meal, dried whey products, and corn distiller's dried solubles. The meal mix for Oregon Pellets consists of herring meal, cottonseed meal, dried whey products, wheat germ meal, shrimp or crab meal, and corn distiller's dried solubles, all in different proportions than for the mash.

The wet mix for the Oregon Mash consists of tuna viscera; turbot, pasteurized salmon viscera, or pasteurized herring; herring or soy oil; kelp meal, and choline chloride.

The wet mix for Oregon Pellets consists of tuna viscera, herring, turbot, salmon viscera, dogfish or hake (herring, salmon viscera, and hake must be pasteurized), kelp meal, herring or soy oil, choline chloride, all in different proportions than for the mash.

The Oregon vitamin premix is the same for the Oregon mash and pellets. It includes ascorbic acid, biotin, B_{12}, E, folic acid, inositol, menadione, niacin, d-pantothenic acid, pyridoxine, riboflavin, and thiamine.

All of the dried meals must pass through a U. S. No. 40 sieve. All wet fish must be ground very fine.

Oregon Starter Mash is formed by thoroughly blending the meal, vitamin, and wet mixes in a manner that will produce granules of a size that will pass 90% or more through a U. S. No. 18 sieve.

The Oregon Pellets are produced in various sizes. The finished pellet must meet sieve test specifications (Table 12). The feed is sacked in paper bags with a net weight of 50 pounds each.

TABLE 12
Pellet Sieve Test Specifications [1]

Pellet size (Inches)	Through U.S. Sieve No.	Over U.S. Sieve No.
$\frac{1}{32}$	16	30
$\frac{3}{64}$	14	20
$\frac{1}{16}$	10	18
$\frac{3}{32}$	7	14
$\frac{1}{8}$	4	8
$\frac{3}{16}$	$\frac{1}{4}$	5

[1] 95% of the sample must meet the sieve test.

The pellets and mash are quick frozen at 20° F or colder and must attain a maximum of 10° F within 24 hours of manufacturing. The feeds are stored in a freezer with a maximum temperature of 0° F.

FEEDING PRACTICES

The primary objective in feeding fish is to feed them adequately with a minimum of waste and the least amount of labor. Fortunately, the feeding habits of salmonids are readily adaptable to large scale feeding practices.

Waste may be reduced and good growth and health maintained by careful attention to food preparation, size and type of food used, frequency of feeding, and close compliance with accepted feeding charts.

The frequency with which fish are fed is governed by the size of the fish and the rapidity with which they clean up the food. When starting fish to feed, it is necessary that small particles of food be available for them to eat. To accomplish this, it is necessary to feed small amounts frequently. As the fish grow, they feed more vigorously and larger amounts of food may be fed at longer intervals.

The amount of food consumed daily is governed by the size of the fish, water temperature, and species. To secure the most efficient food utilization, fish should be fed amounts of dry food considerably less than they will consume.

The growth rate may be increased by feeding all they will eat, but the increased gain rarely compensates for the additional food used. Fish fed all they can eat tend to become sluggish. They do not consume their food quickly and some of the nutrients are lost in the water. Underfeeding results in reduced growth rate and lowered disease resistance. Underfed fish usually have inadequate nutritional reserves to meet unfavorable conditions.

Different amounts of dry food must be fed to rainbow trout at different water temperatures (Table 13). Data presented have been developed from actual production records; however, they have not been tested over the complete range of temperatures, nor are different food values taken into account. The information is the best available at this time and should be followed closely, although minor adjustments may be necessary to meet local conditions.

Feeding Dry Feeds

Ease and rapidity of feeding, less loss of water soluble nutrients, reduction of food preparation time, and use of automatic feeders are some of the advantages of starting trout and salmon fry on dry food or Oregon Moist feed. However, meticulous care and attention are required. If not fed properly, severe losses will

TABLE 13

The Recommended Amount of Dry Food to Feed Rainbow Trout Per Day in Percentage of Body Weight, for Different Size Groups Held in Water of Different Temperatures *

Water temperature in degrees F.	No. fish per pound -2,542	2,542-304	304-88.3	88.3-37.8	37.8-19.7	19.7-11.6	11.6-7.35	7.35-4.94	4.94-3.47	3.47-2.53	2.53-
	Approx. size in inches -1	1-2	2-3	3-4	4-5	5-6	6-7	7-8	8-9	9-10	10-
36	2.7	2.2	1.7	1.3	1.0	0.8	0.7	0.6	0.5	0.5	0.4
37	2.7	2.3	1.8	1.4	1.1	0.9	0.7	0.6	0.5	0.5	0.4
38	2.9	2.4	2.0	1.5	1.2	0.9	0.8	0.7	0.6	0.5	0.5
39	3.0	2.5	2.2	1.7	1.3	0.9	0.8	0.7	0.6	0.6	0.5
40	3.2	2.6	2.2	1.7	1.3	1.0	0.9	0.8	0.7	0.6	0.5
41	3.3	2.8	2.2	1.8	1.4	1.1	0.9	0.8	0.7	0.6	0.5
42	3.5	2.8	2.4	1.8	1.4	1.2	0.9	0.8	0.7	0.6	0.5
43	3.6	3.0	2.5	1.9	1.4	1.2	1.0	0.9	0.8	0.7	0.6
44	3.8	3.1	2.5	2.0	1.5	1.3	1.0	0.9	0.8	0.8	0.6
45	4.0	3.3	2.7	2.1	1.6	1.3	1.1	1.0	0.9	0.8	0.7
46	4.1	3.4	2.8	2.2	1.7	1.4	1.2	1.0	0.9	0.8	0.7
47	4.3	3.6	3.0	2.3	1.7	1.4	1.2	1.0	0.9	0.8	0.7
48	4.5	3.8	3.0	2.4	1.8	1.5	1.3	1.1	1.0	0.9	0.8
49	4.7	3.9	3.2	2.5	1.9	1.5	1.3	1.1	1.0	0.9	0.8
50	5.2	4.3	3.4	2.7	2.0	1.7	1.4	1.2	1.1	1.0	0.9
51	5.4	4.5	3.5	2.8	2.1	1.7	1.5	1.3	1.1	1.0	0.9
52	5.4	4.5	3.6	2.8	2.1	1.7	1.5	1.3	1.1	1.0	0.9
53	5.6	4.7	3.8	2.9	2.2	1.8	1.5	1.3	1.1	1.1	1.0
54	5.8	4.9	3.9	3.0	2.3	1.9	1.6	1.4	1.3	1.1	1.0
55	6.1	5.1	4.2	3.2	2.4	2.0	1.6	1.4	1.3	1.1	1.0
56	6.3	5.3	4.3	3.3	2.5	2.0	1.7	1.5	1.3	1.2	1.0
57	6.7	5.5	4.5	3.5	2.6	2.1	1.8	1.5	1.4	1.2	1.1
58	7.0	5.8	4.8	3.6	2.7	2.2	1.9	1.6	1.4	1.3	1.2
59	7.3	6.0	5.0	3.7	2.8	2.3	1.9	1.7	1.5	1.3	1.2
60	7.5	6.3	5.1	3.9	3.0	2.4	2.0	1.7	1.5	1.4	1.3
61	7.8	6.5	5.3	4.1	3.1	2.5	2.0	1.8	1.6	1.4	1.3
62	8.1	6.7	5.5	4.3	3.2	2.6	2.1	1.8	1.6	1.5	1.4
63	8.4	7.0	5.7	4.5	3.4	2.7	2.1	1.9	1.7	1.5	1.4
64	8.7	7.2	5.9	4.7	3.5	2.8	2.2	1.9	1.7	1.6	1.5
65	9.0	7.5	6.1	4.9	3.6	2.9	2.2	2.0	1.8	1.6	1.5
66	9.3	7.8	6.3	5.1	3.8	3.0	2.3	2.0	1.8	1.6	1.6
67	9.6	9.1	6.6	5.3	3.9	3.1	2.4	2.1	1.9	1.7	1.6
68	9.9	9.4	6.9	5.5	4.0	3.2	2.5	2.1	2.0	1.8	1.7

* The above recommended amounts are given in pounds per 100 pounds of fish.

occur. Feeding should be started when the fry swim up. The smallest size dry food or Oregon Moist feed should be fed at least ten times daily. In order to be sure that food is available most of the time, the fry should be fed slightly more food than is readily eaten. This should be done only during the early feeding stage.

It is very desirable to use automatic feeders (described on the next few pages) in feeding fish, particularly fry, as the time clocks can be set to feed a small amount as many as five times per hour during all of the daylight hours. This keeps feed before the fry almost continually. This is a critical time in their lives to start feeding properly

The food should be placed on the surface of the water as gently as possible, either by hand or with an automatic feeder. Dry food, including the Oregon Moist, fed in this manner will float for several minutes before breaking the surface tension and sinking slowly to the bottom.

As the fish grow, the size of the food should be increased (Table 14). The data are conservative and the fish will take larger food at the upper limits of each size range. Although it is usually better to use food too small rather than too large, delay in increasing the size of the food may lead to gill trouble.

TABLE 14
Size of Dry Food Recommended for Feeding Trout of Various Sizes

Size of food	Size of fish
No. 1	1920–1280 per pound
Nos. 1 and 2 mixed	1360–1120 per pound
No. 2	1200– 800 per pound
Nos. 2 and 3 mixed	880– 480 per pound
No. 3	560– 320 per pound
Nos. 3 and 4 mixed	400– 160 per pound
No. 4	240– 160 per pound
No. 4 and crumbles	240– 80 per pound
Crumbles	160– 48 per pound
Crumbles and 3/32-inch pellets	80– 32 per pound
3/32-inch pellets	32– 9 per pound
5/32-inch pellets	9 per pound and larger

The number of feedings per day may be reduced as the fish grow. Six times daily at 800 fish per pound, four times per day at 400 per pound, and three times at 48 per pound and larger is usually satisfactory.

Fish in raceway ponds may be fed pelleted food by broadcasting the food by hand or with a sugar scoop. Wherever pond design permits, the food should be fed with a mechanical pellet blower. This method does a better job of feeding with considerably less labor.

To prevent waste and obtain maximum growth, regardless of what type food is used, it is necessary to feed trout according to body weight and water temperature. This is especially important when feeding dry food. It is necessary to know the number and total weight of the fish in any trough or pond. A weekly feeding schedule should be followed for each body of water (Figure 61). *Remember that overfeeding will not materially increase growth.* It will increase the conversion rate and food cost and may lead to serious losses.

The Resources Agency of California

Department of Fish and Game

Inland Fisheries Branch

HATCHERY FEEDING SCHEDULE

Hatchery_____ Week of_____ 19___, through_____ 19___ Water temp._____

Pond	Variety	Average size	Number	Weight	Daily feed requirements			Losses	Remarks
					Lbs.	Size	Per-cent		

FG 705 (Rev. 6/71)

FIGURE 61—Hatchery feeding schedule.

TABLE 15

Amount and Types of Dry Food Needed to Raise 25,000 Rainbow Trout From Swimup to Six Per Pound

Size of fish	Length	Total weight in pounds at start of period	Total weight in pounds at end of period	Pounds gained	Percentage of body weight to be fed	Pounds of food to be fed per day	Pounds of food needed	Type of food to be used*	To reach size	Days
				Water Temperature 56 Degrees F.						
3200/lb.	0.80	7.8	19.5	11.7	7.0	1.4	21.0	No. 1 fry food	1280/lb.	15
1280	1.25	19.5	52.1	32.6	6.0	3.1	46.5	No. 2 fry food	480	"
480	1.75	52.1	111.6	59.5	5.3	5.8	87.0	No. 3 fry food	224	"
224	2.25	111.6	195.3	83.7	4.6	9.0	135.0	No. 4 fry food	128	"
128	2.75	195.3	347.2	151.9	4.3	14.9	223.5	Crumbles	72	"
72	3.25	347.2	520.8	173.6	4.0	27.1	406.5	Crumbles	48	"
48	3.75	520.8	781.2	260.4	3.6	28.1	421.5	Crumbles	32	"
32	4.25	781.2	1,116.1	334.9	3.3	36.8	552.0	3/32 in. pellets / Crumbles	22.4	"
22.4	4.75	1,116.1	1,506.0	389.9	2.7	40.7	610.5	3/32 in. pellets	16.6	"
16.6	5.25	1,506.0	1,923.0	417.0	2.4	46.2	693.0	3/32 in. pellets	13	"
13	5.75	1,923.0	2,500.0	577.0	2.3	57.5	862.5	3/32 in. pellets	10	"
10	6.25	2,500.0	3,125.0	625.0	2.2	68.8	1,032.0	3/32 in. pellets	8	"
8.5	6.75	3,125.0	3,846.2	721.2	2.1	80.8	1,212.0	3/32 in. pellets	6.5	"
6.5	7.25	3,846.2	4,545.4	699.2	2.0	90.1	1,351.5	3/32 in. pellets	5.5	"
Total				4,537.6			7,654.55			

Conversion = 1.68

Does not provide for losses during period.

Requirements

```
No. 1 fry food ------     21.0 lbs.
No. 2 fry food ------     46.5 lbs.
No. 3 fry food ------     87.0 lbs.
No. 4 fry food ------    135.0 lbs.
Crumbles ------------  1,116.75 lbs.
3/32 in. pellets ----  6,248.25 lbs.

Total --------------  7,654.55 lbs.
```

TABLE 15—Continued

Amount and Types of Dry Food Needed to Raise 25,000 Rainbow Trout From Swimup to Six Per Pound

Water Temperature 51 Degrees F.

Size of fish	Length	Total weight in pounds at start of period	Total weight in pounds at end of period	Pounds gained	Percentage of body weight to be fed	Pounds of food to be fed per day	Pounds of food needed	Type of food to be used*	To reach size	Days
3200/lb.	0.80	7.8	19.5	11.7	6.0	1.2	24.0	No. 1 fry food	1280/lb.	20
1280	1.25	19.5	52.1	32.6	5.0	2.6	52.0	No. 2 fry food	480	"
480	1.75	52.1	111.6	59.5	4.3	4.8	96.0	No. 3 fry food	224	"
224	2.25	111.6	195.3	83.7	3.7	7.2	144.0	No. 4 fry food	128	"
128	2.75	195.3	347.2	151.9	3.3	11.5	230.0	Crumbles	72	"
72	3.25	347.2	520.8	173.6	3.0	15.6	312.0	Crumbles	48	"
48	3.75	520.8	781.2	260.4	2.8	21.9	438.0	Crumbles	32	"
32	4.25	781.2	1,116.1	334.9	2.6	29.0	580.0	Crumbles	22.4	"
22.4	4.75	1,116.1	1,506.0	389.9	2.4	36.1	722.0	3/32 in. pellets	16.6	"
16.6	5.25	1,506.0	1,923.0	417.0	2.2	42.3	846.0	3/32 in. pellets	13	"
13	5.75	1,923.0	2,500.0	577.0	2.0	50.0	1,000.0	3/32 in. pellets	10	"
10	6.25	2,500.0	3,125.0	625.0	1.9	59.4	1,188.0	3/32 in. pellets	8	"
8	6.75	3,125.0	3,846.2	721.2	1.8	69.2	1,384.0	3/32 in. pellets	6.5	"
6.5	7.25	3,846.2	4,545.4	699.2	1.7	77.3	1,546.0	3/32 in. pellets	5.5	"
Total				4,537.6			8,562.0			

Conversion = 1.88

*Does not provide for losses during period.

Requirements

No. 1 fry food	——	24.0 lbs.
No. 2 fry food	——	52.0 lbs.
No. 3 fry food	——	96.0 lbs.
No. 4 fry food	——	144.0 lbs.
Crumbles	——	1,051.0 lbs.
3/32 in. pellets	——	7,195.0 lbs.
Total	——	8,562.0 lbs.

TABLE 15—Continued

Amount and Types of Dry Food Needed to Raise 25,000 Rainbow Trout From Swimup to Six Per Pound

Water Temperature 46 Degrees F.

Size of fish	Length	Total weight in pounds at start of period	Total weight in pounds at end of period	Pounds gained	Percentage of body weight to be fed	Pounds of food to be fed per day	Pounds of food needed	Type of food to be used*	To reach size	Days
320)/lb.	0.80	7.8	19.5	11.7	5.0	1.0	30.0	No. 1 fry food	1280/lb.	30
128)	1.25	19.5	52.1	32.6	4.5	2.3	69.0	No. 2 fry food	480	"
480	1.75	52.1	111.6	59.5	4.1	4.6	138.0	No. 3 fry food	224	"
224	2.25	111.6	195.3	83.7	3.7	7.2	216.0	No. 4 fry food	128	"
128	2.75	195.3	347.2	151.9	3.3	11.5	345.0	Crumbles	72	"
72	3.25	347.2	520.8	173.6	2.9	15.1	453.0	Crumbles	48	"
48	3.75	520.8	781.2	260.4	2.5	19.5	585.0	Crumbles	32	"
32	4.25	781.2	1,116.1	334.9	2.2	24.5	735.0	3/32 in. pellets	22.4	"
22.4	4.75	1,116.1	1,506.0	389.9	2.0	30.1	903.0	3/32 in. pellets	16.6	"
16.6	5.25	1,506.0	1,923.0	417.0	1.8	34.6	1,038.0	3/32 in. pellets	13	"
13	5.75	1,923.0	2,500.0	577.0	1.6	40.0	1,200.0	3/32 in. pellets	10	"
10	6.25	2,500.0	3,125.0	625.0	1.5	46.9	1,407.0	3/32 in. pellets	8	"
8	6.75	3,125.0	3,846.1	721.1	1.4	53.8	1,614.0	3/32 in. pellets	6.5	"
6.5	7.25	3,846.1	4,545.4	699.3	1.3	59.1	1,773.0	3/32 in. pellets	5.5	"
Total				4,537.6			10,506.0			

Conversion = 2.3

Does not provide for losses during period.

Requirements

No. 1 fry food	30.0 lbs.
No. 2 fry food	69.0 lbs.
No. 3 fry food	138.0 lbs.
No. 4 fry food	216.0 lbs.
Crumbles	1,458.0 lbs.
3/32 in. pellets	8,595.0 lbs.
Total	10,506.0 lbs.

* Size based on type of food presently produced by reliable fish feed companies.

Sample weight counts should be made at least every 2 weeks but it is preferable to make them weekly, particularly if the water is over 55° F and the fish are growing rapidly. The feed schedule should be adjusted each time the weight counts are made. This will make a more efficient operation.

A

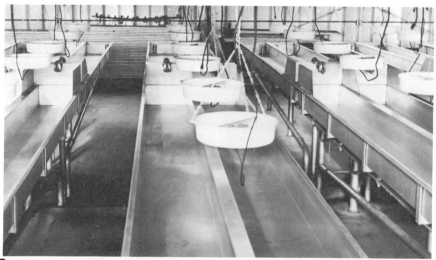

B

FIGURE 62—**A**, Allen automatic feeders are suspended over hatchery troughs and operated on a time clock to feed fish at frequent intervals; **B**, feeders are suspended over the center divider of a pair of troughs so each feeder supplies feed to two troughs. (Note wiper blade inside the pans). *Photographs by George Bruley, 1973.*

Improper feeding technique can lead to serious trouble. One of the most common feeding faults is the tendency to feed the ponds rather than the fish. It is no harder to feed the fish. Care in feeding is always well rewarded.

The total amount of dry food of various sizes required to raise 25,000 rainbow trout from swimup to six fish per pound will vary with water temperature (Table 15). Data presented may be slightly altered to meet local conditions, and should be helpful in estimating dry food requirements.

Allen Automatic Feeder

Allen Feeders can be used on hatchery troughs and various types of tanks and ponds (Figure 62). The pan containing feed has holes of several sizes for the feed to drop through. All holes are closed with tape except the one of proper size to allow the correct amount of feed to drop into the water. Two wiper blades with a clearance of $\frac{1}{16}$ inch rotate in the pan to push the feed over the holes. The blades are driven by an electric motor of $\frac{1}{250}$ hp and makes $4\frac{3}{4}$ rpm.

Hatchery troughs are in pairs so a feeder can drop feed into two troughs simultaneously by hanging it over the center of the paired troughs. Two or three feeders are used on a pair of troughs 16 feet long.

The Allen Feeder can be supported in many ways over various types of tanks and ponds. It is important to have the feed distributed well when the fry are starting to feed as they will not move far to seek food. The operation of the feeders is controlled by time clocks so small amounts of feed can be disbursed at very frequent intervals during the daylight hours and allows the fish to be fed during hours other than the normal working day.

Neilsen Automatic Feeder

The Neilsen feeder is adaptable for use on concrete ponds (Figure 63), and can also be used on various other types of ponds. This type feeder has a basic capacity of 12 pounds of feed. An adapter can be put on the feeder to increase its capacity to 50 pounds.

A fan at the bottom of the feeder is propelled at high speed by an electric motor. When the motor is activated, feed is released through an adjustable opening to the fan which broadcasts it over a large area. The operation of these feeders is controlled by time clocks to provide small amounts of food at frequent intervals. Fish become accustomed to the location of the feeding stations and spend enough time in the area to feed well and make rapid growth. Automatic feeders can be adjusted so there will be no waste of food, although it is necessary to change the adjustments frequently as the fish grow and require more food. These feeders are insulated so Oregon Moist feed can be used as well as the regular dry food.

Mechanical Pellet Blower

Several designs of mechanical pellet blowers mounted on various kinds of mobile vehicles have been constructed and used to broadcast pelleted food into raceway ponds or other accessible ponds. The blower (Figure 64) has a storage bin divided into four compartments, so different sizes of food may be carried as needed. The operator releases food from any compartment desired into a common hopper in which the food is weighed so the operator can control the correct amount of food scheduled for a single pond or a group of ponds.

A

B

FIGURE 63—Neilsen automatic feeder. **A**, feeder in operation. Feed can be seen dropping from a pipe at bottom of feeder onto a fan that broadcasts it over the pond. **B**, feeder is fastened to the concrete wall so it can be turned 180 degrees to hang on the outside of the wall when the mechanical crowder is in operation on the pond walls. *Photographs by George Bruley, 1973.*

FIGURE 64—Mechanical pellet blowers can be mounted on a mobile vehicle or on a trailer. Some models have a common hopper under the four food storage compartments in which the food for a pond or group of ponds can be weighed before broadcasting into the ponds. The rate and amount of food released is controlled from the driver's compartment. *Photograph by George Bruley, 1973.*

A gasoline engine drives a blower that forces air through the blower tube at a high rate. The operator releases food from the common hopper into the blower tube from which it is broadcast into the pond. It is quite important that the operator watch the density of fish in the various areas of the pond and release the food at the proper rate so it will not be wasted. He can also change the rate of speed of the vehicle on which the feeder is mounted.

This pellet feeder is loaded quite easily by gravity flow of pellets from feed storage bins, and is an efficient method of feeding pond fish. It was manufactured by Neilsen Metal Industries Inc., 3501 Portland Road, N. E., Salem, Oregon 97303.

GRADING FISH

It has been said that a hungry trout 3 inches long will devour a trout 1½ inches long, providing he can catch it, and that a 6 inch trout will eat a 3 inch trout, and so on up the line, until the word cannibalism becomes frightening. There just isn't any question that large trout will eat small trout. Both trout and salmon are handled in comparatively large numbers when reared in troughs or ponds. Therefore, any lot of fish of fair numbers must consist of the offspring of several females. This, plus the fact that all of the eggs from a single female are not always of the same size, and the fact that the larger fish in any group are better able to compete for food than the smaller fish, explains some of the reasons why trout and salmon grow irregularly in size.

Grading of fish is necessary to get good growth, reduce cannibalism, prevent competition between fish of smaller size with their larger kin, and obtain fish of the correct size to meet management requirements. Furthermore, the total weight of fish in any one group can be more accurately determined for computing the amount of food to feed in percentage of body weight, if the fish are of a nearly

A

B

FIGURE 65—Wilco adjustable grader. **A,** grader bars are in the open position which can be adjusted by turning the knob on top of the grader and locking it with the thumb nut; **B,** grader bars are closed; note the position of the knob on top of the grader; graders can be made any length required. *Photographs by George Bruley, 1973.*

even size. Grading equipment has been designed which will grade fish to a number of sizes in one operation. Grading in more sizes than is necessary, however, is time consuming and normally grading one group of fish into three sizes—small, medium, and large—is sufficient.

Many types of fish grading devices are currently in use for segregating the various sizes of fish from the time they hatch until they are ready for release in the stream. Each of these devices has been developed to operate under a certain set of conditions to accomplish a desired objective. All of them are efficient to some degree and can be adapted to almost all situations, with varying degrees of efficiency.

If fish are raised in hatchery troughs to the fingerling size, they are generally graded before putting them into rearing ponds. In most situations, the fish are graded one or two times while they are in the rearing ponds. It is general practice to grade fish before planting which can be done very easily with the fish pump.

Fish graders most commonly used in California are the Wilco adjustable fish grader or modification, and the Fish Pump Grader; the latter being the more efficient.

Wilco Adjustable Grader

The Wilco adjustable fish screen or grader (Figure 65) was developed by Wilco Products, 327 Burnett Ave., North, Renton, Washington 98055. It can be used to grade fish in a concrete pond or the concrete trunk of an earthen pond.

The grader consists of oval shaped vertical bars in a frame $61\frac{3}{4}$ inches long, 30 inches high, and $1\frac{1}{2}$ inches thick. Openings between the vertical bars can be adjusted by turning the adjusting knob, located near the end of the screen, to the desired setting and locking it by tightening a thumb nut. The openings can be varied from $\frac{1}{16}$ inch minimum to $\frac{13}{16}$ inch maximum. Larger screens or graders are available by special order.

The grader is placed in the upstream end of the pond trunk of earthen ponds. The fish are pulled into the trunk behind the grader with a seine and crowded against the grader with a push rack. The mechanical crowder described earlier, or a hand push rack, can be used on ponds with concrete walls much more efficiently than using a seine. The push rack on the mechanical crowder is a Wilco design. It can be used for grading or herding any size fish (Figure 66). As the fish become crowded between the racks, those small enough to pass through the grader swim into the pond above. Those unable to pass through remain in the pond behind the grader where they can be temporarily held for planting or other purposes.

The flow and water depth of the pond should be held at normal operating levels during the grading operation. The time required to grade a pond of fish varies from only a few minutes to 2 hours depending somewhat on local conditions. Several graders can be operated at one time in the same pond series should the need arise.

The advantage of the Wilco grader is that it is almost automatic. The only operation required is that of selecting the size of spacing, setting the grader in place, and forcing the fish in behind it. The attendant can then continue with other duties and within a short time the fish have graded themselves with a minimum of handling.

FIGURE 66—The rack attached to the mechanical grader is a Wilco grader which can be used for grading fish or crowding them together, depending on the setting of the grader. (Note fish in the pond ahead of the grader). *Photograph by George Bruley, 1973.*

Murray-Hume Automatic Grader

The Murray-Hume automatic grader can be made at a hatchery. It is a rack similar to the Wilco grader with vertical aluminium tubes equally spaced for the size of fish to be graded. Several racks should be available with different spacings to grade different size fish. The operation is the same as described for the Wilco grader. The advantage of the Wilco grader is that the vertical bars can readily be adjusted to any space required.

There is a serious disadvantage to this method of grading fish with the Wilco or Murray-Hume graders in that from 10 to 25% of the small fish do not pass out through the vertical bar graders. This method is definitely not as efficient as other methods of grading fish although it is less time consuming.

Fish Pump

The 6 inch fish pump (Figure 67) has many uses as a labor saving piece of equipment in grading fish, moving fish between ponds, and loading fish into tank trucks.

The pump unit used is a Paco Horizontal Non-clog Pump, Type NCH, directly connected with a gas engine. It has a single port, non-clog impeller with the blade edge well rounded to avoid catching trash, or in this case, injuring fish. This type pump is used in conveying fruits, vegetables, and food products in food processing plants. It is manufactured and distributed by Pacific Pumping Company in Oakland and Los Angeles.

The pump is mounted on a trailer that can be moved to the ponds where it is to be used. It has a reinforced suction hose attached to a pickup box that is placed in a section of the pond from which the fish are to be removed. The pickup box has a gate that is perforated so it will draw water at all times that the pump is running, and fish only when it is opened. The fish pass through the pump up to

FIGURE 67—Fish pump loading a tank truck. Fish are sucked out of the pond, pass through the pump, pushed up to the separator tower and spilled onto a grader rack which slopes downward. The large fish slide across the grader rack and go into the tank truck. The small fish and water drop through the rack and return to a pond via a hose. *Photograph by George Bruley, 1973.*

the separator tower and a grader box equipped with an adjustable Neilsen grader that can be set from $\frac{1}{8}$ inch minimum to 1 inch maximum spacings which slope downward so the fish will slide into a chute and drop into a tank truck. The water and small fish drop through the parallel bars and pass through a hose back into a pond. The separator tower can be raised or lowered hydraulically to accommodate different size tank trucks. This pump can load fish into a tank truck in a small fraction of the time required by any other method. It is only necessary to be sure the fish are crowded into the area of the suction hose and pickup box.

Fish can be moved from one pond to another by removing the parallel bar grader so all water and fish will pass through the discharge hose and be piped to any pond within a reasonable distance.

Fish can be graded by pumping them onto the parallel bar grader with proper spacings for the size fish desired. The graders can be changed easily to grade different size fish. A modification of the Wilco adjustable grader could be designed so one adjustable rack could be used instead of the several interchangeable racks. The smaller fish pass through the grader and discharge hose with the water and can be piped to a pond in the nearby area. The larger fish can be loaded into a tank truck and hauled to another pond or run through another pipe to another pond by supplying water to this outlet.

FIGURE 68—Spring balance scale suspended over hatchery trough for weighing fingerlings. *Photograph by J.H. Wales, 1959.*

The fish pump grader does a more thorough job of removing the small fish than can be accomplished with the Wilco Adjustable Grader method previously described. Small fish do not have much chance of getting into a tank truck load of fish for planting if the grader in the pump is adjusted properly. This is the most rapid and efficient method of grading catchable and subcatchable size fish.

This fish pump has been assembled by Neilsen Metal Industries, Inc., 3501 Portland Road, N. E., Salem, Oregon 97303.

A 3 inch fish pump has been designed for use with fingerlings. It is the same principle as the 6 inch pump but has a lighter suction attachment like a vacuum sweeper so it can be moved around easily to catch the smaller fish.

WEIGHING AND ENUMERATING FISH

A fish hatchery, like any other business, must account to its stockholders in terms of number, size, and weight of fish produced, as well as maintain a running inventory of fish on hand for proper feeding and management. Any system of accounting is time consuming and costly and, in the case of hatcheries, the difficulties are increased by the fact that the system must be based on the determination of size of millions of small fish that are delicate, very active, and must be handled with care. Three sizes of fish are produced and distributed by California hatcheries: fingerlings, subcatchables, and catchables. The number of fish per pound, regardless of size, is determined with a suspended spring balance scale (Figure 68). The number of fingerlings in a shipment is determined with the same scale. Subcatchables and catchables may be weighed by the scale method or the displacement method, employing a fish pump (Figure 67) and a tank truck with sight gauge (Figure 69). Instructions for weighing and recording fish for transfer or shipment follow.

Fingerlings—Scale Method

When fish average one or more to the ounce, or 16 or more to the pound, they are classed as fingerlings. The weight and all calculations of these small fish are in ounces for greater accuracy. The weights are then converted to pounds for all records (See Table 16 for speedy conversion).

A suspended type of spring balance scale is used in weighing all fingerlings.

The dial of the scale is divided into 80 ounces and records 15 pounds in three revolutions of the pointer. Any support used for suspending the scale over the trough or pond should be readily portable and of a type that leaves the working area entirely clear, with no overhanging projections.

A 3 gallon lightweight bucket should be used. After the support and scales are set up, the bucket is hung on the scales and filled with water until the pointer comes to about 70 on the first revolution. Enough water is added with a dipper to bring the pointer to zero. This leaves a capacity of 160 ounces for weighing fish.

Before starting the count it is necessary to determine by actual hand count the number of fish per ounce. This is the sample upon which calculations of the number of fish are based. The sample count should be very accurate. In order to get a random sample the fish should be concentrated, which makes the fish easier to catch and provides a more accurate sample.

To make the sample count, fish are poured from the dip net into the bucket until a predetermined number of ounces is reached.

TABLE 16
Fish Per Ounce to Fish Per Pound

No/oz	No/lb	No/oz	No/lb	No/oz	No/lb
1	16	51	816	101	1,616
2	32	52	832	102	1,632
3	48	53	848	103	1,648
4	64	54	864	104	1,664
5	80	55	880	105	1,680
6	96	56	896	106	1,696
7	112	57	912	107	1,712
8	128	58	928	108	1,728
9	144	59	944	109	1,744
10	160	60	960	110	1,760
11	176	61	976	111	1,776
12	192	62	992	112	1,792
13	208	63	1,008	113	1,808
14	224	64	1,024	114	1,824
15	240	65	1,040	115	1,840
16	256	66	1,056	116	1,856
17	272	67	1,072	117	1,872
18	288	68	1,088	118	1,888
19	304	69	1,104	119	1,904
20	320	70	1,120	120	1,920
21	336	71	1,136	121	1,936
22	352	72	1,152	122	1,952
23	368	73	1,168	123	1,968
24	384	74	1,184	124	1,984
25	400	75	1,200	125	2,000
26	416	76	1,216	126	2,016
27	432	77	1,232	127	2,032
28	448	78	1,248	128	2,048
29	464	79	1,264	129	2,064
30	480	80	1,280	130	2,080
31	496	81	1,296	131	2,096
32	512	82	1,312	132	2,112
33	528	83	1,328	133	2,128
34	544	84	1,344	134	2,144
35	560	85	1,360	135	2,160
36	576	86	1,376	136	2,176
37	592	87	1,392	137	2,192
38	608	88	1,408	138	2,208
39	624	89	1,424	139	2,224
40	640	90	1,440	140	2,240
41	656	91	1,456	141	2,256
42	672	92	1,472	142	2,272
43	688	93	1,488	143	2,288
44	704	94	1,504	144	2,304
45	720	95	1,520	145	2,320
46	736	96	1,536	146	2,336
47	752	97	1,552	147	2,352
48	768	98	1,568	148	2,368
49	784	99	1,584	149	2,384
50	800	100	1,600	150	2,400

TABLE 16—Continued

Fish Per Ounce to Fish Per Pound

No/oz	No/lb	No/oz	No/lb	No/oz	No/lb
151	2,416	201	3,216	251	4,016
152	2,432	202	3,232	252	4,032
153	2,448	203	3,248	253	4,048
154	2,464	204	3,264	254	4,064
155	2,480	205	3,280	255	4,080
156	2,496	206	3,296	256	4,096
157	2,512	207	3,312	257	4,112
158	2,528	208	3,328	258	4,128
159	2,544	209	3,344	259	4,144
160	2,560	210	3,360	260	4,160
161	2,576	211	3,376	261	4,176
162	2,592	212	3,392	262	4,192
163	2,608	213	3,408	263	4,208
164	2,624	214	3,424	264	4,224
165	2,640	215	3,440	265	4,240
166	2,656	216	3,456	266	4,256
167	2,672	217	3,472	267	4,272
168	2,688	218	3,488	268	4,288
169	2,704	219	3,504	269	4,304
170	2,720	220	3,520	270	4,320
171	2,736	221	3,536	271	4,336
172	2,752	222	3,552	272	4,352
173	2,768	223	3,568	273	4,368
174	2,784	224	3,584	274	4,384
175	2,800	225	3,600	275	4,400
176	2,816	226	3,616	276	4,416
177	2,832	227	3,632	277	4,432
178	2,848	228	3,648	278	4,448
179	2,864	229	3,664	279	4,464
180	2,880	230	3,680	280	4,480
181	2,896	231	3,696	281	4,496
182	2,912	232	3,712	282	4,512
183	2,928	233	3,728	283	4,528
184	2,944	234	3,744	284	4,544
185	2,960	235	3,760	285	4,560
186	2,976	236	3,776	286	4,576
187	2,992	237	3,792	287	4,592
188	3,008	238	3,808	288	4,608
189	3,024	239	3,824	289	4,624
190	3,040	240	3,840	290	4,640
191	3,056	241	3,856	291	4,656
192	3,072	242	3,872	292	4,672
193	3,088	243	3,888	293	4,688
194	3,104	244	3,904	294	4,704
195	3,120	245	3,920	295	4,720
196	3,136	246	3,936	296	4,736
197	3,152	247	3,952	297	4,752
198	3,168	248	3,968	298	4,768
199	3,184	249	3,984	299	4,784
200	3,200	250	4,000	300	4,800

To determine the number of fish per ounce, the fish are counted from the bucket and the number of fish counted is divided by their weight in ounces.

Example: Ten ounces of fish are counted and found to number 160. Number of fish (160) divided by the number of ounces (10) equals number of fish per ounce (16), or:

$$10 \quad / \quad \frac{16 \text{ fish per ounce}}{160}$$

In counting into cans, the number of ounces weighed into each can is multiplied by the number of fish per ounce. This number multiplied by the number of cans equals the total number of fish.

Subcatchables and Catchables—Scale Method

When fish average less than 16 per pound and more than six per pound, they are classed as subcatchables. Fish weighing six per pound and larger fish are classed as catchables. The weight of both classes is recorded in pounds.

A suspended type of spring balance scale of 60 pound capacity is used in determining the number of fish per pound.

The dial of the scale is divided into ounces and pounds and records 60 pounds in three revolutions of the pointer.

A 6 gallon bucket should be used. Water is added until the pointer makes one revolution to zero. This leaves a capacity of 40 pounds for weighing fish. Ordinarily, the fish should be weighed in increments of 10 or 20 pounds to facilitate calculation.

The procedure is the same as the one used in enumerating fingerlings, except that pounds are used instead of ounces.

When weighing fish, it is necessary that an accurate count be kept of the number of buckets weighed and placed in cans. To keep count of the number of buckets weighed without the use of some mechanical aid is not practical. One of the most satisfactory procedures is to use a hand tally register (see specifications). The hand tally register can, in most instances, be fastened to the scale support within easy reach of the person doing the weighing. Thus, each bucket of fish weighed can be tallied as it is taken from the scales.

Scale Specifications

Scales, for weighing *fingerlings:* "Chatillion", hanging, graduated in ounces, 80 ounces per revolution, capacity 240 ounces, 8 inch dial.

Scales, for weighing *subcatchables* and *catchables:* "Chatillion", hanging, graduated in ounces, 20 pounds per revolution, 60 pound capacity, 7 inch dial.

Bucket, for weighing *subcatchables* and *catchables:* 6 gallon, 24 gauge steel, with bail handle, no lid, straight sides, unpainted or treated, type normally used for paint containers.

Counter hand tally: Braun-Knecht Heimann Company No. 23661, tallies from 1 to 10,000.

Displacement Method

The displacement method of weighing fish is based on the weight of water displaced by a pound of fish. This method is used when transporting fish larger than 16 per pound in tanks.

Specific gravity tests have shown that in the size range from one to sixteen per pound an average of 1.018 pounds of trout displace 1 pound of water. The standard deviation in these tests was 0.0102.

The figure 1.018 was rounded off to 1.02. Therefore, the total pounds of water displaced multiplied by 1.02 equals the pounds of trout loaded.

Example: 1,100 pounds of water $\times$ 1.02 = 1,122 pounds of trout.

All that is needed to convert a planting tank to this method is a sight gauge mounted near the top of the tank. This consists of an 18 inch length of ½ inch glass boiler water gauge mounted vertically on the tank beside a strip of brass channel. It is important that the tube and the brass channel be exactly parallel (Figure 69).

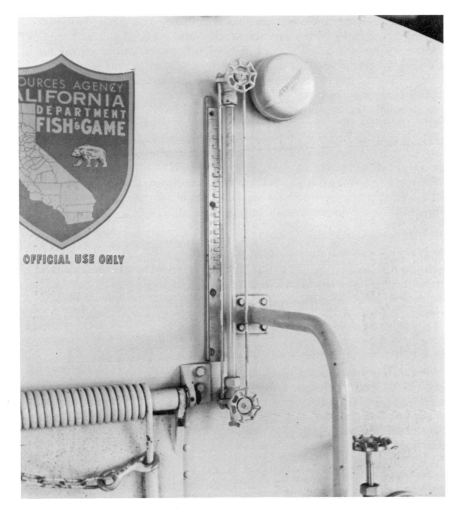

FIGURE 69—Glass sight gauge with calibrated metal marker strip on back of fish planting tank for weighing fish by the displacement method when the fish pump is used for loading. *Photograph by George Bruley, 1973.*

The first step in calibrating the gauge is to fill the tank with water. The water is drawn down to the top of the sight gauge and the level marked on the brass channel with a pencil. A combination square placed across the channel with the blade extended to the glass tube helps make the mark accurate.

The water is drawn off into a tub, set on a platform scale, in specified increments: 100, 50, and 25 pounds. The channel is marked at the level of each increment.

Care must be taken to guard against splashing. Any water accidentally drawn off in excess of the increment should be returned to the tank before marking the channel.

The importance of extreme accuracy in this operation cannot be overemphasized. all the fish loaded into the tank will be measured by the gauge.

When the desired number of marks has been made on the brass channel, it is removed and placed in a vice and the pencil marks are scribed into it.

Before loading the fish, the truck is filled with water well above the mark indicating the size of the load desired and driven to the loading location.

Surplus water is drawn off until the water is exactly even with the starting mark. It is not necessary that the tank be level. However, if the truck is moved during loading, the amount of fish loaded should be noted and a new starting point selected for the remainder of the load.

Weight samples are taken in the usual manner to determine the size of the fish being loaded, the number of fish per load being determined by multiplying the number of fish per pound by the number of pounds of fish loaded.

When loading with a fish pump, the truck is positioned, the water is drawn off to the selected point, and loading is begun. At hatcheries not equipped with a fish pump, the fish are dipped directly into the tank.

LENGTH-WEIGHT RELATIONSHIP IN RAINBOW TROUT
Condition Factor

The ratio of length to weight in trout is often expressed as the condition factor. It is an index to the weight of a fish in relation to its length and is the yardstick often applied when checking growth rate or determining the amount of food to be fed at fish hatcheries.

The factor is obtained from the following equation:

$$C.F. = \frac{W \times 100,000,}{L^3}$$

in which C.F. equals the coefficient of condition in the English system, W is the weight of the fish in pounds, L is the length in inches (usually fork or total length), and 100,000 equals the factor which places the decimal point, usually at the right of two significant figures.

Length-weight relationships for California hatchery rainbow trout have been determined (Table 17).

TRANSPORTATION OF TROUT AND SALMON
Tank Trucks

Fish transport tank trucks are of various sizes and design (Figures 70, 71, 72, 73 and 74). California uses five sizes of tanks for fish distribution: 2500, 1200, 600, 400, and 150 gallon. All tanks are insulated so temperatures can be held more constant. The three larger tanks (2500, 1200, and 600) have refrigeration units; ice is used for temperature control in the smaller units. The newer tank trucks

TABLE 17
Number of Rainbow Trout Per Unit of Weight by Length

Fork length in inches	Number fish per pound	Fork length in inches	Number fish per pound
1.1–1.3	1040–1600	8.0– 8.2	4.39
1.4–1.6	560–1024	8.3– 8.5	3.38
1.7–1.9	320– 544	8.6– 8.8	3.35
2.0–2.2	192– 304	8.9– 9.1	3.25
2.3–2.5	140.8	9.2– 9.4	3.15
2.6–2.8	108.8	9.5– 9.7	2.80
2.9–3.1	86.4	9.8–10.0	2.53
3.2–3.4	63.36	10.1–10.3	2.34
3.5–3.7	49.6	10.4–10.6	2.18
3.8–4.0	37.28	10.7–10.9	2.02
4.1–4.3	29.76	11.0–11.2	1.89
4.4–4.6	24.96	11.3–11.5	1.75
4.7–4.9	20.8	11.6–11.8	1.60
5.0–5.2	17.6	11.9–12.2	1.45
5.3–5.5	15.28	13	1.13
5.6–5.8	12.74	14	0.91
5.9–6.1	10.84	15	0.73
6.2–6.4	9.52	16	0.61
6.5–6.7	8.25	17	0.50
6.8–7.0	7.51	18	0.42
7.1–7.3	6.30	19	0.36
7.4–7.6	5.35	20	0.31
7.7–7.9	4.46		

are equipped with a generator so the refrigerators and water circulating pumps can be run by electricity.

The pumps and refrigerators are driven by separate gasoline engines on other models. The aeration system is generally designed with the water drawn from the bottom of the tank by pumps. It passes through a venturi to inject air into the water and is then sprayed back into the tank over refrigeration coils. The 400 gallon tanks have small electric pumps at each end of the tank which operate from a heavy duty truck battery. The water is picked up at the bottom of the tank and

FIGURE 70—A 2500 gallon fish transport truck. The water pumps and refrigeration units are operated by electricity which is generated by a diesel driven unit on the truck. *Photograph by George Bruley, 1973.*

FIGURE 71—A 1200 gallon fish transport truck. The water pumps and refrigeration units are powered with gasoline engines. *Photograph by George Bruley, 1973.*

sprayed back. This type of tank holds the temperature very constant without refrigeration because the water does not pass through a venturi on the outside of the tank and draw warm air from the atmosphere into the water. Generally, the 150 gallon tanks have a gasoline driven pump which draws the water from the bottom of the tank, passes it through a venturi and sprays it into the tank. Some tanks are also supplied with bubbles of oxygen through a carborundum stone from a bottle of compressed oxygen. The fish planting equipment, particularly the larger tank trucks, are complicated and expensive. There should be an operating manual with each piece of equipment.

FIGURE 72—A 600 gallon fish transport truck. Electric water pumps and refrigeration units are powered by a gasoline generator. *Photograph by George Bruley, 1973.*

FIGURE 73—A 400 gallon fish planting tank is mounted on a flatbed pickup truck. It is equipped with a small electric pump at each end of the tank which operates from a generator or a heavy duty truck battery and sprays the water to the middle of the tank. It holds the temperature very constant. *Photograph by George Bruley, 1973.*

FIGURE 74—Several 150 gallon fish planting tanks are mounted on pickup trucks. The pumps are gasoline driven. Ice is used for refrigeration. *Photograph by George Bruley, 1973.*

The efficiency of the fish transport tank trucks has been improved through the past years to increase the hauling capacity, although it does vary throughout the state due to different conditions.

Size Tank	Catchable Size Trout
2500 gallon	4000–5000 pounds
1200 gallon	2000–2500 pounds
600 gallon	1000–1200 pounds
400 gallon	800–1000 pounds
150 gallon	350–400 pounds

The water used in planting tanks should be clean and preferably from the hatchery water supply. Foreign matter in the water could clog the circulating system and reduce its efficiency. Foul water from a previous load of fish could remain in the pump chambers and other parts of the system in sufficient amounts to affect the following load adversely. For this reason it is essential always to flush out planting tanks and the circulating system before filling the tank for the next load.

Disinfection of Planting Tanks

As a precautionary measure against planting equipment carrying a disease from a planted water back to the hatchery, all planting tanks used off hatchery grounds should be disinfected as soon after each plant as possible and before reuse, unless there is no contact between the planting tank and the planted water. Nets, buckets, pipes, boots, etc., which make contact with the planted water should not be allowed to infect the planting tank and should be disinfected before reuse.

The disinfection of planting tanks may be accomplished by using ½ ounce (dry weight) of HTH (70% available chlorine) per 25 gallons of water for 30 minutes.

In waters which have a pH above 6, it will be necessary to add 1 ounce (fluid ounce) of glacial acetic acid per 100 gallons of water.

To disinfect tanks and other implements with HTH (chlorine) the proper procedure is:

1. Fill the tank or other container with the desired amount of water.
2. Add the proper amount of acetic acid to the water.
3. After the water and acetic acid are mixed, add the required amount of HTH to the acidified water. DO NOT ADD THE ACID TO THE DRY HTH.

In the case of spray tanks, it will be possible to sterilize the unit by filling the tank to one-half capacity and running the pumps and spray for the desired exposure period. Thorough flushing of the tank and pumps is required before reuse.

Caution should be exercised in disposal of treatment solution as the chlorine level may still be toxic to fish.

To reach required concentration for the following capacity tank, use the amounts as follows:

Tank	HTH	Glacial acetic acid
150 gallon	3 ounces	1.5 ounces
600 gallon	12 ounces	6 ounces
1200 gallon	24 ounces	12 ounces
2500 gallon	50 ounces	25 ounces

Utility Tank

A small tank (Figure 75) about 3 feet square and 2 feet deep with a capacity of about 100 gallons has been developed with a drain pipe similar to a regular fish planting tank. The aeration system on this type of tank is compressed oxygen from a bottle regulated through a cylinder type carbon stone. This type tank can be picked up with a fork lift and used to move small fish in the hatchery operation. It is particularly effective when swimup fry are removed from the incubator trays and hauled to various ponds.

FIGURE 75—This small 100 gallon utility tank is used on the hatchery grounds to move fish. It can be moved about with a fork lift. *Photograph by George Bruley, 1973.*

Airplane Fish Planting

Tremendous savings can be realized by planting fish in inaccessible lakes by airplane instead of by the old method of packing them on horses and mules. California plants fingerling trout in several hundred inaccessible mountain lakes annually. Fingerlings can be dropped from 300 to 1000 feet into a lake without injury. The elevation varies according to type of terrain and various other conditions. Successful plants can be made in lakes as small as two acres.

A twin engine Beechcraft has been equipped to haul 26 5-gallon aluminum cans with fish (Figure 76). An electrically driven air compressor supplies each can with bubbles of air through hoses and carborundum stones. As the airplane

A

B

FIGURE 76—**A**, loading airplane with 5 gallon aluminum cans containing fish to be planted; **B**, air is supplied to each can in the airplane through manifold and hoses on right side from an electrically driven air compressor in center of picture; hopper from which fish are dropped is at right of picture. *Photographs A by Edwin P. Pister and B by George Bruley, 1973.*

approaches the lake to be planted, the number of cans of fish for planting are poured into a hopper which is over a hole in the bottom of the plane. At the proper time, all fish and water are released instantaneously through a trap door.

At this writing four tanks are being designed with eight compartments each, for a total of 32 compartments. Each compartment will have a capacity to carry the same amount of water and fish previously hauled in a 5 gallon can. Each tank with eight compartments will have a hopper to drop the fish from the number of compartments desired in any particular lake. The fish will be loaded in the compartments at the airport. The copilot will be able to release the fish from any number of compartments desired and from the hopper by pushing buttons. No one need be in attendance in the rear of the plane.

Can Trucks

Fingerling fish are transported to the airport for aerial planting in 5 gallon aluminum cans, 11 inches in diameter and 19 inches high, on various size trucks. Usually the trucks are equipped with gasoline engines driving a compressor to supply air through hoses and carborundum stones to each can of fish. The most desirable truck for this purpose in California is a refrigerated van truck capable of hauling 64 5 gallon cans or two flights on the airplane (Figure 77). Plans are under way to remodel the van so that 96 5 gallon cans could be hauled to make three flights when the tanks for the plane which will haul the equivalent of 32 cans are constructed. Flat bed pickup trucks are also equipped to haul up to 64 5 gallon fish cans or two flights on the plane.

Effects of Metabolic Waste Products on Fish During Transportation

Not a great deal is known about the effects of waste products of metabolism on trout and salmon in pond management and in fish transportation. It is known, however, that ammonia, urea, uric acid, carbon dioxide, and other products of metabolism can create a toxic condition. When the concentration of ammonia and other excretory products increases in the water, trout lose the ability to use oxygen and the blood picture changes drastically. For example, as the ammonia concentration increases to 1 ppm, the oxygen concentration in the blood decreases to about $\frac{1}{7}$ normal, and the carbon dioxide content increases about 15%, with resulting suffocation.

Carbon dioxide may be one of the major limiting factors in fish transportation. It has been shown that (a) whenever carbon dioxide is kept below 15 ppm with reasonable values of oxygen and temperature, the fish remain in good condition, and (b) when carbon dioxide reaches about 25 ppm, the fish become distressed.

Measures which may reduce the accumulation of metabolic wastes are:

1. *Starvation of fish before shipment.* Fish should not be fed for 24 to 48 hours before transportation. Common practice is to keep them off all feed for 24 hours, but a longer period is desirable for long hauls, especially if the hatchery water is below 50° F.

2. *Maintenance of low temperatures.* The most suitable temperature range for transporting fish in tanks is from 47 to 53° F.

3. *Use of hypnotic drugs.*

4. *Removal of metabolic products.* This may be accomplished by aeration or the addition of buffering agents. Lime water has been satisfactorily used as a buffering agent to control carbon dioxide in experimental tests. However, this is not practical under operating conditions. Further experiments along this line are necessary.

A

B

FIGURE 77—**A**, refrigerated van truck is desirable to haul cans of fish to the airplane because the temperature remains more nearly constant; **B**, hoses can be seen which supply air to the cans from air compressors driven by gasoline engines. *Photographs A by T. T. Jackson and B by George Bruley, 1973.*

Sodium Amytal in Fish Transportation

Sodium amytal, one of the many hypnotic barbiturates available, produces a slow reacting, long lasting tranquilizing effect on trout and salmon. Some speculation remains as to why sodium amytal is an aid in fish transportation. Two theories have been advanced along this line; one is that the drug slows down the activity of the fish, thereby reducing its oxygen requirements, and the other is that sodium amytal slows down the entire catabolic process, with a general reduction of metabolic wastes which affect oxygen consumption in fish.

The concentration at which sodium amytal has been used in fish transportation varies considerably. The most satisfactory dosage is generally ½ grain per gallon of water. The effectiveness of the drug is, to some extent, regulated by temperature. It appears to decrease when the temperature rises above 54° F, the most effective range being from 47 to 54° F.

Sodium amytal was used in transporting fish in tank trucks when its benefits were first discovered. This increased the carrying capacity a great deal. In recent years, the aeration systems on the tank trucks have been improved to the extent that the use of sodium amytal does not increase the number of fish that can be carried. It is now rarely used in tank truck transportation of fish in California except in the large tanks for maximum loads.

Sodium amytal is used in transporting fingerlings in cans by truck and on the airplane which increases the carrying capacity considerably. The aeration system of adding bubbles of air or even bottled oxygen is apparently not as efficient as the systems on the tank trucks; therefore, the use of sodium amytal can still increase the carrying capacity in some instances.

Use of Antifoam Emulsion

The formation of foam and scum, especially when drugs are used in transporting fish, often becomes quite bothersome. This is especially true when long hauls of heavy loads are being made. Foam on the surface of the water makes it difficult to observe the fish being carried. Foam and scum will also dry on the tank surfaces and must be cleaned off. The formation of foam and scum can be prevented by the use of Dow Corning Antifoam AF Emulsion. The Antifoam AF Emulsion is available as a 30% concentration of the Dow Corning Antifoam A, in water and emulsifiers.

Diluting the emulsion to a 10% solution in warm water (9 parts water, 1 part emulsion), makes it easy to measure and handle. For dilutions of the AF Emulsion, use any convenient measuring vessel.

Maximum effectiveness of the emulsion can best be realized by adding the Antifoam AF Emulsion before adding either drug or fish. When introduced into the tank the emulsion may impart a slightly turbid or opaque quality to the water. This nontoxic additive has a wide latitude of effectiveness and is very easy to use. Foaming can be controlled by adding Antifoam Emulsion, one ounce or 25 cc. of the 10% dilution for every 100 gallons of water.

Since sodium amytal is not commonly used in transporting fish in tank trucks, the foam and scum problem has decreased. Consequently, the advantages of using Antifoam Emulsion are not as great, but its use does keep the water more clear so the fish can be observed better. It is still commonly used in tank truck transportation of fish.

TROUT AND SALMON DISEASES

This section refers primarily to specific fish diseases which principally concern fish culturists in California. It should be borne in mind that there are others which are extremely important in other states.

Some diseases occur at certain hatcheries at regular intervals. When fish culturists are aware of this, there often is sufficient forewarning to treat the fish before the disease reaches a serious stage.

Symptoms

When organisms become numerous on a fish, they may cause changes in its behavior or produce other obvious symptoms. Unfortunately, each disease or parasite does not always produce a single symptom or syndrome characteristic in itself. Nevertheless, by observing the symptoms one can usually narrow down the cause of the trouble.

Some of the obvious changes in behavior of fish suffering from a disease, parasite, or other physical affliction are: (1) loss of appetite, (2) abnormal distribution in pond, such as riding the surface, gathering at the pond sides or in slack water, and crowding the head or tail screens, (3) flashing, scraping on bottom or projecting objects, darting, whirling, or twisting, and loss of equilibrium, and (4) loss of vitality, weakness, and loss of ability to stand handling during grading, seining, loading, or transportation.

In addition to changes in behavior, disease may produce physical symptoms, or the parasite may be seen by the unaided eye. For microscopic examination, it is necessary to call in a fish disease expert. Symptoms observed may be external or internal, or a combination of both.

Gross external symptoms are: (1) discolored areas on the body, (2) eroded areas or sores on the surface of the body, head, and fins, (3) swelling on the body or gills, (4) popeye, (5) hemorrhages, and (6) cysts containing parasites.

Gross internal symptoms are: (1) color changes of organs or tissue (pale liver or kidney or congested organs), (2) hemorrhages in organs or other tisues, (3) swollen or boil-like lesions, (4) change in texture of organs or tissues, (5) accumulated fluid in body cavities, and (6) cysts containing parasites.

Diseases and Parasites

Before taking up specific diseases, it will be best to describe briefly a general classification of disease-causing organisms among fish. The majority of such organisms may be placed in two groups. One is the plant kingdom. Bacteria and fungi belong to this group. The other is the animal kingdom. Forms belonging to this group vary from the relatively simple, single celled animals (Protozoa) to the complex, multicelled organisms (Metazoa) such as worms, copepods, and mussels.

In addition, there are two other classes of diseases important to fish culturists. One is nutritional in nature, the other is caused by virus. As previously mentioned, diseases or parasites may be found externally or internally, or both in some cases.

External Bacterial Diseases

Columnaris

Columnaris disease is caused by *Chondrococcus columnaris* and is known to many fish culturists. It is not restricted to salmonids, but is also found on many warm water fishes. This organism, although studied by a number of investigators, remains confusing to specialists. Because of details in its life history, some workers refer to the organism as *Cytophaga columnaris.* This disease is most prevalent in California when water temperatures rise above 56° F. It affects fish ranging in size from small fingerlings to large catchables. The early symptoms are grayish-white areas on the body, head, fins, or gills. On careful observation, the edges of these lesions may appear reddened and hemorrhaged. The early lesions may start at a weakened or injured area and spread. In some advanced cases the fish may be almost entirely covered by the lesions. The lesions are often invaded secondarily by fungus. At this stage, these areas appear fuzzy or furry.

Most workers consider this disease to be both internal and external in nature, though only the external lesions are seen. In order to control this infection, it is usual to treat with copper sulfate flushes for one to three successive days for the external stage and to feed Terramycin at the rate of 36 grams of TM-50 or TM-50 D per 100 pounds of fish (4 grams of pure antibiotic per 100 pounds of fish) for 10 consecutive days. In the past, sulfamerazine (8 grams/100 pounds of fish) was used but has generally been replaced by Terramycin.

When water temperatures are high and fish are crowded, the disease will recur; under such conditions the treatment only controls the disease and does not eliminate it. In order to keep columnaris in check, it is necessary to give repeated treatments.

Bacterial Gill Disease

This disease is one of the most common in California, and year in and year out probably causes more losses than any other. Rainbow trout, as well as salmon, may be infected. The bacterium responsible for most cases of bacterial gill disease is probably an unidentified species related to *Chondrococcus columnaris.* The conditions under which this organism flourishes are water temperatures above 56° F and crowding of the fish.

In early stages the gills may be swollen and clubbed, with large amounts of mucus present. The fish may be weak, crowd the pond tail screens, ride high in the water, and "go off feed." In some cases there is much destruction of the gill tissue. The necrotic tissue in these cases is grayish-white and many of the gill filaments may be completely eroded. Quite often the gills on one side only are affected.

Generally in California, dipping or flushing with copper sulfate affords control. This is probably due to the fact that the organism is restricted to the gills in most cases. On occasion when the external treatment (copper sulfate) has not yielded the desired results, Terramycin (36 grams TM-50 or TM-50 D/100 pounds of fish) may be used as with Columnaris.

Peduncle Disease

This disease (Figure 78) is probably the same as those known as cold water or low temperature disease. In California this affliction, while not common, can cause serious mortalities when it occurs. It is usually found among fingerling rainbow trout but also infects catchable size rainbow. Salmon fingerlings (King and Silver) are also susceptible. The organism responsible for this disease is believed to be *Cytophaga psychrophila* and is found at temperatures below 56° F as well as at higher temperatures; and contrary to reports from other states, we find the fish do not need to be crowded for it to appear.

The tissue of the caudal fin and peduncle is affected. In early stages it may involve only the tail; later, it involves the peduncle causing much tissue destruction. Advanced stages result in complete loss of the caudal fin and the posterior end of the peduncle leaving exposed muscle and bone of the vertebral column. Occasionally the peduncle is well involved with the caudal fin being unaffected. Combination treatments of copper sulfate dips or flushes with sulfamerazine (8 grams per 100 pounds of fish) or Terramycin (36 grams TM-50 or TM-50 D/100 pounds of fish) added to the diet will control the disease, though sometimes no treatment will avail.

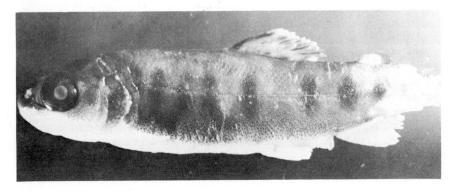

FIGURE 78—Peduncle disease is sometimes a serious problem encountered when water temperatures are in the 40°–50°F range. *Photograph by W.E. Schafer, 1965.*

Fin Rot

Fin rot is a symptom that appears to be related to several possible agents. Bacteria, crowded conditions, concrete ponds, or improper diets are some of the reported causes. Often in bacterial cases the early stages exhibit a white discoloration along the outer edge of fins. As the disease advances, this moves toward the base of the fins. The tissues including fin rays are destroyed, although the rays are lost more slowly and usually appear as ragged remnants. Quite often when crowded, rainbow trout will develop a dorsal fin rot which superficially resembles bacterial fin rot. It shows a smooth white thickened fin margin and lacks the ragged appearance and eroded fin rays of the bacterial form. Dipping infected fish in a 1:2000 solution of copper sulfate for 1–2 minutes is usually effective in controlling the bacterial form of the disease. Overcrowded conditions and improper diets (low folic acid or high or excess Vitamin A) require other corrective approaches. Most species of salmonids are susceptible to this disease.

Internal Bacterial Diseases

Furunculosis

This bacterial disease, caused by *Aeromonas salmonicida,* has appeared in records of fish disease since 1894 and is believed to be limited to freshwater and anadromous fishes.

This disease has caused severe mortalities in hatchery and wild trout. Rainbow trout are generally considered to be among the more resistant to furunculos, but they do get the disease and can provide a source of infection for other fish. In the past, this disease was one of the most feared because no adequate treatments were known. With the advent of the sulfa drugs, however, a major step forward was taken, and at the present time there is little excuse for encountering the hatchery losses which once resulted from furunculosis.

In yearling and older fish, the disease is usually marked by a series of open sores on the body. Usually one can find a combination of open sores and raised boil-like lesions. The disease is essentially a "blood poisoning". The bacteria are carried about in the blood. They may collect in clumps in the smaller blood vessels, then rupture the blood vessels and invade the surrounding tissues.

The lesions and sores produced may appear to the unaided eye as swollen red spots beneath the skin. As the disease progresses these spots may fuse, destroying the tissue and enlarging into a definite swollen area. These areas may break open on the outside of the fish, forming good-sized ulcers. Sometimes the bacteria cause soft and blister-like lesions filled with blood. The gills may also show hemorrhaged areas.

There may be a marked congestion of blood vessels in the body cavity. The lining of the intestine may be inflamed and there may be a discharge of blood and mucus from the vent, especially after death. The spleen may be enlarged and be a bright cherry red; this is particularly true in fingerling trout. The kidney is usually badly diseased and may be converted to a semiliquid mass. Trout may die before advanced sores and boils form. In fingerlings, frequently, the only evident symptoms are dark irregular areas on the sides between the dorsal and pectoral fins, which may also become affected and be reduced to nothing more than the rays.

Sulfamerazine (10 grams/100 pounds of fish per day) in the diet has been an old standby in the treatment of furunculosis, though in recent years, because of sulfa resistant strains, Terramycin (36 grams TM-50 or TM-50 D/100 pounds of fish per day for 10 days) has become the drug of choice to feed. However, it is best to routinely run sensitivity tests on all furunculosis isolates as some have been found to be resistant to Terramycin. Furox 50 has been used successfully under experimental conditions against sulfamerazine and Terramycin resistant isolates of *Aeromonas salmonicida.*

Ulcer Disease

As the name implies, this disease, caused by *Hemophilus piscium,* is characterized by ulcers or sores on the surface of the fish. These resemble furunculosis but are essentially different in that the "sores" begin on the outside and work through the skin, whereas in furunculosis they develop beneath the skin as blood filled boils and may eventually break open on the outside if the fish lives long enough. Ulcer disease "sores" are often circular in outline, though they may be irregular. Another important characteristic of the disease is the frequent infection of the jaws and roof of the mouth. Positive diagnosis of ulcer disease requires procedures usually available only in a well developed bacteriological laboratory. This disease has not been reported in western North America.

Recommended treatments are the use of chloramphenicol or terramycin in the food at the rate of 2½ to 3½ grams of pure antibiotic activity per 100 pounds of fish per day until losses have dropped to "normal."

Red-mouth Disease

This disease is generally thought to be caused by a specific enteric bacteria which has been isolated but not characterized by generic classification. It is also considered to be restricted to rainbow trout.

Unfortunately, several other identified bacteria have been isolated from rainbow and other salmonids which produce similar or overlapping symptoms. One of the most frequently encountered organisms is *Aeromonas liquefaciens.* The symptoms of red-mouth include lethargy, darkness, reddened skin lining the mouth and throat, hemorrhages are found on the operculum, isthmus, gills, and associated membranes as well as the base of the fins. Internally, inflammation and hemorrhages are often seen in the posterior intestine. The visceral fat, liver, and lining of the body cavity are often studded with small hemorrhages. These symptoms may not always be present and it may be necessary to examine the blood microscopically and carry out appropriate laboratory procedures. Both the red-mouth organism and *Aeromonas liquefaciens* are usually controlled by a standard Terramycin feeding of 36 grams of TM-50 (or TM-50 D) per 100 pounds of fish daily for 10 days. Sulfamerazine fed at 10 grams per 100 pounds of fish per day for 10 days has also been used with reasonable success. Terramycin has usually been the drug of choice when bacterial sensitivity tests have been carried out.

Fish Tuberculosis

For many years organisms related to the human tuberculosis bacterium have been reported from fish. Extensive work has been done in the Pacific Northwest especially on salmon. In California the disease has been found in adult king salmon, silver salmon, and steelhead. The disease has been minimized in hatchery fish by eliminating the practice of feeding infected salmon viscera and carcasses. One outbreak among 80,000 fingerling silver salmon was traced to a commercially prepared diet which contained untreated salmon viscera. Approximately 25% of these fish were infected. There are no drugs presently known that can be used to treat this disease among salmonids. However, the practice of not feeding raw or untreated salmon viscera in the diet can reduce the incidence of the disease among hatchery reared salmonids to a rare occurrence among returning adult fish.

In affected fish the disease resembles miliary tuberculosis in humans. The lesions are caseous or purulent tubercles scattered among the kidney, liver, spleen, and digestive tract. The bacteria in active cases may be found in virtually all tissues. The significance of this disease among the andromous salmonid fishery is unknown.

Kidney Disease

This disease has not been a problem of significance in California. In the early 1930's a serious outbreak was recorded among some eastern brook trout in a southern California commercial fish hatchery. Since then only an occasional case has been found and usually among eastern brook trout, although juvenile silver and king salmon have been found with the typical symptoms. Small gram positive organisms have been seen among some rainbow trout brood fish but without typical symptoms. This bacterial disease has been reported to cause serious losses among trout in the eastern United States and in juvenile salmon reared in Pacific Northwest hatcheries. The gross symptoms include popeye, skin hemorrhages, blebs, hemorrhages at the bases of the fins and sometimes deep ulcers which can be seen externally. The abdomen may be greatly swollen. Internally, the kidney, liver, and spleen often exhibit "white, boil-like" lesions similar to those seen in fish tuberculosis. The body cavity often contains fluid. In the past, most of the cases among salmon were attributed to feeding of raw or untreated infected salmon viscera or carcasses.

Some outbreaks have been related to wild infected fish in the water supply. Circumstantially some workers believe the disease may be transferred on, or in fertilized eggs.

Treatments with sulfonamides (sulfamerazine and sulfamethazine) at approximately 10 grams/100 pounds of fish per day for 10 days may reduce the loss rate. This level is considered therapeutic and if removed from the diet, the losses will rise again. When a prophylactic level of 2 grams/100 pounds of fish per day is fed continually, following the therapeutic treatment, the losses are minimized until liberation of the fish. Erthromycin at 4.5 grams/100 pounds of fish per day for 3 weeks has been used successfully. Terramycin at its standard dose (36 grams TM-50 or TM-50 D/100 pounds of fish) has also been used.

External Protozoan Diseases

Trichodina spp.

These ciliated protozoans (Figure 79) parasitize most salmonids in California. When numerous, they can cause serious losses among fingerling trout and salmon as well as among larger fish. *Trichodina* is found on the body surface, fins, and gills. Infected fish "flash" in an attempt to scratch off the offending organisms. Irregular whitish areas of a superficial nature appear on the body and the fins may become frayed. Scales may loosen and congested areas appear on the skin. *Trichodina* may be controlled by 1:500 acetic acid dip for 1 minute or a 1:4000 or 1:6000 formaldehyde bath for 1 hour.

Epistylis spp.

These stalked ciliates (Figure 80) are among the most common protozoans on fish in California. They are found most frequently on fingerling rainbow trout. A light infestation is usually of no consequence. When present in large numbers, these protozoans irritate the fish, causing them to flash. At this stage, 1:500 acetic acid dip or a 1:4,000 formaldehyde bath for 1 hour will control the organism.

Chilodonella cyprini

These ciliated protozoans (Figure 81) are not often found on fish in California hatcheries. On the few occasions when they were observed, they were present in numbers great enough to cause losses, "flashing," and a slight cloudy appearance of the body among small king salmon and steelhead fingerlings. The same, or a very similar, parasite has been observed on warmwater fish. In all cases effective control resulted by a 1:500 dip of acetic acid for 1 minute. More than likely a 1:4000 or 1:6000 1 hour bath with formaldehyde would provide similar control.

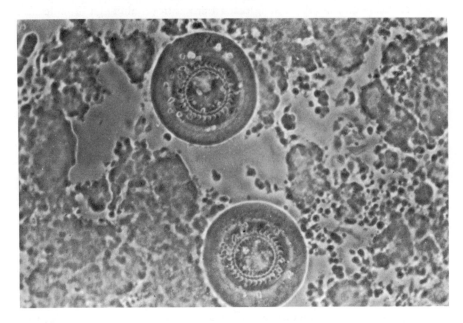

FIGURE 79—*Trichodina* spp. from a rainbow trout skin scraping. *Photograph by W.E. Schafer, 1963.*

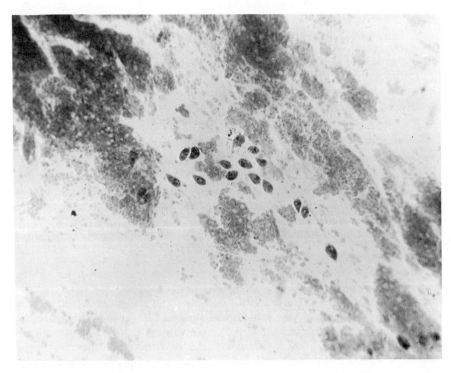

FIGURE 80—*Epistylis* sp. on skin of rainbow trout. *Photograph by Harold Wolf, 1958.*

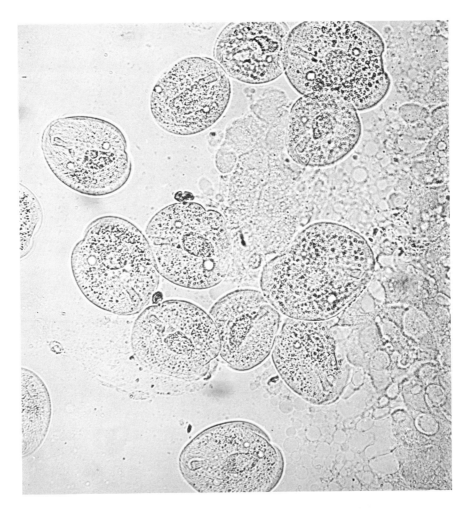

FIGURE 81—*Chilodonella cyprini* an external protozoan parasite of trout, salmon, and warm water fishes. *Photograph by W.E. Schafer, 1965.*

Costia spp.

There are probably two species of this parasite which can be found on all salmonid species reared in California hatcheries (Figure 82). They are *Costia necatrix* and *C. pyriformis* with the former believed to be found only on the body surface, whereas *C. pyriformis* is believed to occur on the gills as well as on the body. Both species, if not controlled, can cause severe losses especially among young or small fish. Larger fish that are heavily infected may suffer losses and often come down with additional diseases.

Affected fish may flash, be listless, seek slack water, crowd the bottom, back to the screens, and lose their appetite. Often a bluish or greyish film will be noted on the body. A microscopic examination is needed for a positive identification of this parasite. A 1 minute dip in 1:500 acetic acid or a 1:4000 or 1:6000 1 hour formaldehyde bath is usually very effective. Sometimes a follow-up treatment is necessary to clean up the infestation.

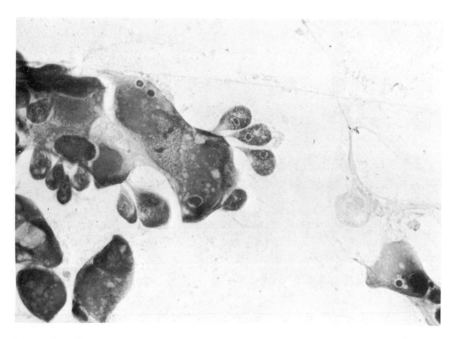

FIGURE 82—*Costia* spp. attached to tissue from skin scraping of a cutthroat trout. *Photograph by Harold Wolf, 1972.*

FIGURE 83—"Ich", *Ichthyophthirius multifilliis,* on skin of rainbow trout; a moderate infestation. *Photograph by J. H. Wales, 1958.*

Ichthyophthirius multifiliis

"Ich" (Figure 83), as this ciliated protozoan is familiarly known to fish culturists, is found on almost all kinds of freshwater fishes. It has on occasion caused serious losses. Warm water and crowding are conducive to the outbreak of "Ich."

Fish infested with this parasite may flash, crowd the water inlet, or later seek slack water areas. When observed closely, the infected fish reveal small, white swellings on the body surface, which may appear roundish. If there are many parasites they may be crowded together sufficiently to present irregular raised patches. When single parasites are observed on the gills, they may appear more oval shaped than round. Positive identification requires a microscopic examination.

FIGURE 84—Parasitic copepods *Salmincola edwardsii* attached to gill filaments of a rainbow trout. *Photograph by W.E. Schafer, 1967.*

"Ich" has a complicated life cycle. The free swimming stage may be controlled by a 1:4,000 1 hour formaldehyde treatment. The parasitic stage, which is imbedded in the superficial layer of the skin, resists any known treatment which does not kill the fish. Because the stages on the fish mature and leave the fish at various intervals, it would be necessary to treat for the free swimming stages at such frequent intervals that this method is impractical, from the standpoint of the fish's ability to stand repeated treatments, and because of the time involved.

One of the best methods for controlling this parasite is to place the infected fish in shallow, swiftly moving water, not allowing the dead fish to accumulate, and sweeping the pond bottoms daily if conditions permit. This method removes the parasites as they leave the fish. The encysted form may remain on the fish for a period of 10 days to 5 weeks, depending on the water temperature. The warmer the water, the shorter it remains on the fish. If clean fish are to be put into a pond which contained fish that had the "Ich," it is best to allow the pond to dry out for several days. If this is not possible, sterilization of the pond with formaldehyde should suffice. Recently Malachite green flushes (3–6 ounces of dry malachite green in 5–10 gallons of water per 2–3 cfs flow) have been used against "Ichs" as a daily or every other day routine with success. The treatment is usually continued (along with the low water regime) for approximately 2 weeks depending on the individual outbreak.

Miscellaneous External Parasites

Copepods

Two copepod genera are of importance in California: *Salmincola* and *Lernaea.*

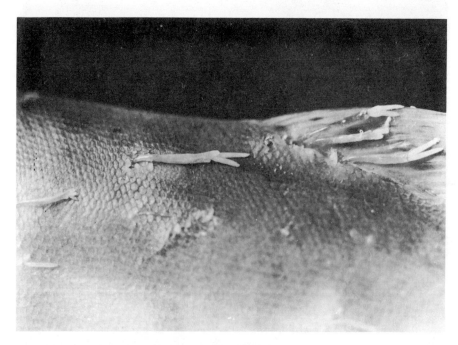

FIGURE 85—The parasitic copepod *Lernaea carassii* commonly known as the anchor parasite shown attached to fish. *Photograph by W.E. Schafer, 1967.*

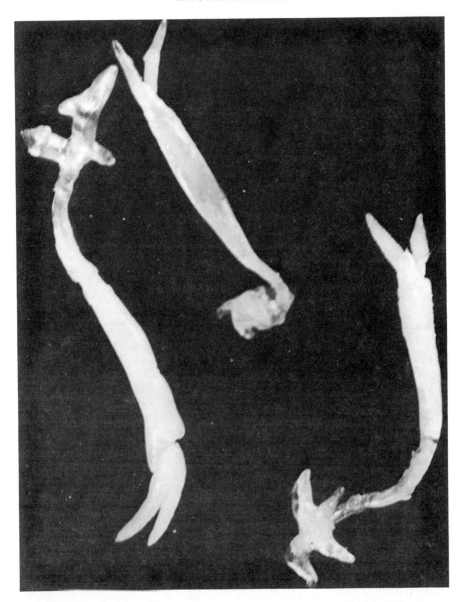

FIGURE 86—Parasitic copepods *Lernaea carassii* removed from fish and showing the anterior (usually embedded) end that earns them the name "anchor" parasites. *Photograph by W.E. Schafer, 1967.*

Salmincola edwardsii

This copepod (Figure 84) has been found mainly on rainbow trout and steelhead. Only on occasion has it been found on other species of salmonids. This parasite has a complicated life cycle. Adults are found on the gills and axillary regions for the most part. Sometimes they may also be found in the mouth. A heavy infestation debilitates the fish, provides a route for secondary infections, and when on the gills makes repiration more difficult. Eggs may be seen attached to the adult copepod or are sometimes found free in the water with free-swimming larvae. The free swimming larvae become permanently attached to a fish and mature; the cycle is then repeated.

The adult is very resistant to chemicals, and to date no method which will kill it without also killing the fish has been developed. The free swimming stages are reported to be killed by strong salt solutions, or by using 1:6,000 formaldehyde for 1 hour. Since the adult copepod may remain alive on a fish for 2 months or more, treatments to kill free swimming stages are not effective. Partial control may be effected by keeping infected fish in swiftly moving water.

Lernaea carassii

This copepod, commonly called "anchor worm" (Figures 85 and 86), has not been a problem in California hatcheries, although it is seen on a great variety of wild fish. It may occur at the base of the fins or scattered about the body surface. Occasionally it penetrates the eye and causes blindness. It is likely that the methods of controlling *Salmincola* would be effective in controlling *Lernaea.*

Parasitic Worm

Gyrodactylus elegans

"Gyros," as they are commonly known, are small worms found on all species of salmonids in California hatcheries (Figure 87). They seem to exert their most serious damage on fingerling and catchable-size rainbow trout. The heaviest infestations usually occur in winter and early spring when the water temperatures are in the 40–50° F. range. Severe infestations, if not controlled properly, are often followed by peduncle disease or *columnaris.* Successful treatments for "Gyros" are acetic acid dips (1:500 for 1 minute), or formaldehyde baths (1:4000 or 1:6000 for 1 hour). Controlling "Gyros" in cold water (40–50° F.) is difficult and may require repeated treatments to prevent them from getting out of control and then having bacterial diseases superimposing. The best approach is to diagnose the early presence of the parasites and keep them from becoming abundant by appropriately timed treatments.

Fungus

Saprolegnia parasitica

Several genera of fungi have been reported to attack fish and fish eggs. *Saprolegnia parasitica* is the species of importance in California. *Saprolegnia* is generally a secondary invader. Injuries or wounds caused by external parasites usually provide the initial site for infection. The fungus usually presents a grayish-white, furry appearance. Fish affected with fungus may be dipped in a 1: 15,000 malachite green solution for 10 to 60 seconds with beneficial results. Salmonid eggs, when threatened by fungus, may also be treated with malachite green as described in the section on treatments.

Internal Protozoan Diseases

Hexamitus salmonis

This flagellated protozoan (Figure 88), best known to fish culturists as octomitus, is found in the intestines of trout and salmon. This organism is not believed to cause serious losses. In fact, the opinion is growing that its presence does not indicate a serious condition.

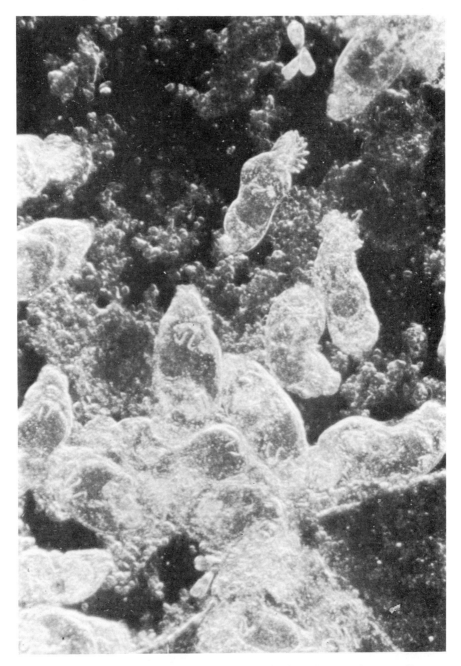

FIGURE 87—A group of external parasitic worms *Gyrodactylus elegans* from a rainbow trout skin scraping. *Photograph by Don Manzer, 1972.*

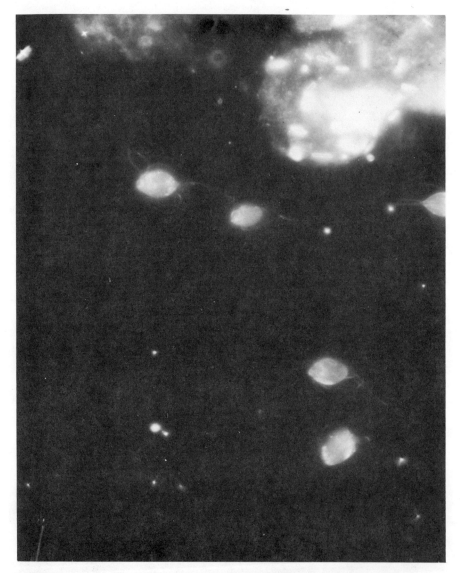

FIGURE 88—*Hexamitus salmonis (Octomitus)*, as seen in the intestinal contents under dark field observation. *Photograph by Harold Wolf, 1958.*

It may be found in healthy, well-formed fingerlings as well as in thin or pinheaded ones. The only sure method of determining the presence of *Hexamitus* is by microscopic examination of the intestinal contents. No treatment is considered necessary for controlling this organism.

Cryptobia borreli

Cryptobia borreli, a blood-inhabiting flagellate (Figure 89), has been found in California in salmonids; suckers, *Castostomus,* and sculpins, *Cottus.* The protozoan is transmitted from one fish to another by leeches. This parasite has caused considerable trouble among rainbow trout brood fish. The kidneys' function is upset and infected fish often are anemic, develop popeye, and contain fluid in the body cavity.

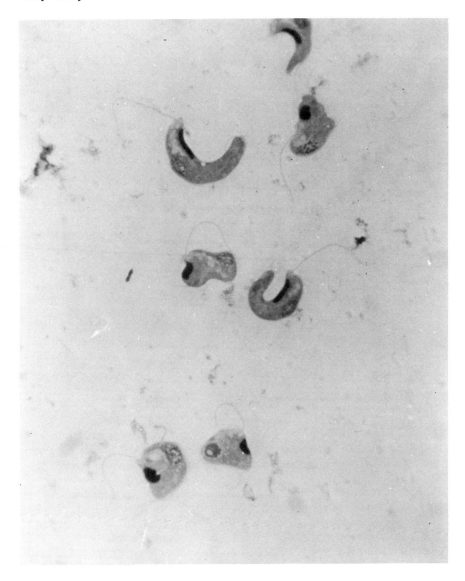

FIGURE 89—*Cryptobia borreli,* a flagellated protozoan found in the blood of salmonids and other fish. *Photograph by Harold Wolf, 1958.*

Positive diagnosis requires microscopic examination and identification of the protozoan. In very heavy infections the protozoan may be found not only in the blood but beneath the scales and between the muscle segments. No satisfactory method for treating this parasite is known at the present time.

Ceratomyxa shasta

Myxosporidian protozoa are common parasites of fish, probably causing little damage or few mortalities with most species. *Ceratomyxa shasta* (Figure 90), however, has caused very serious losses among rainbow trout and steelhead in one California hatchery. This parasite may invade virtually every tissue in the fish's body, causing great damage to the tissues.

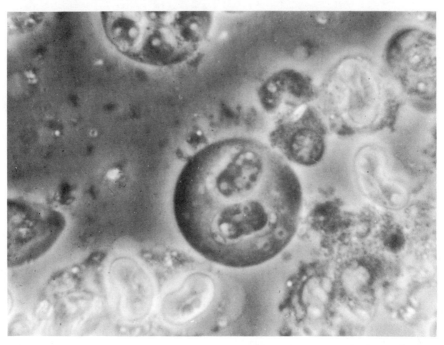

FIGURE 90—The myxosporidian protozoan *Ceratomyxa shasta.* Developing stages and mature spores found in ascitic fluid in a rainbow trout. *Photograph by W.E. Schafer, 1963.*

In other states, this parasite has been a significant problem among adult king and silver salmon. In California, only occasionally has it been observed in adults of these species. Under some circumstances, juvenile king and silver salmon have been found infected.

With most parasites in this group, transmission is believed to be from fish to fish by ingestion of the spore which is thought to be the infective unit. This has not been established to be the case with *Ceratomyxa*. Mt. Shasta strains of rainbow trout are believed to be very susceptible and the Pit River strain of rainbow has a high degree of resistance to the infection. Other species of trout (brook, brown, Eagle Lake) are more resistant than the Mt. Shasta rainbow. Steelhead from the Klamath and Eel River drainages are quite susceptible and suffer severe losses when exposed. The disease in California appears to be restricted to certain drainages, usually those that involve a lake or impounded water in the system. When a susceptible species is exposed in such a water supply and the water temperature is above 50° F., the fish will show evidence of the disease in approximately 38 days. The incubation period shortens as the water temperature increases.

There are no known drugs or treatments at this time which will control this parasite. This leaves resistant strains of trout as the most promising method of carrying out successful fish planting in infected waters.

Plistophora salmonae

This microsporidian parasite forms white appearing cysts on the gills of rainbow trout, king, and silver salmon in California. It has also been observed on the gills of sculpins, *Cottus* sp. When the parasite is abundant, it has been associated with mortalities among rainbow trout and king salmon. There are no known treatments at this time.

Miscellaneous Internal Parasites

Blood Fluke

Sanguinicola davisi

This trematode worm on occasion has contributed to serious losses among hatchery reared rainbow (including Kamloops), steelhead and cutthroat trout. Its distribution in California is coincidental with the snail *Oxytrema circumlineata* which serves as the intermediate host for the worms. The adult worm lives in the main blood vessels (arteries) leading from the heart to the gills.

The mature worm lays eggs which lodge in the gill capillaries. The eggs develop to a miracidia which bores its way out of the gills. It is at this time that it causes most of its damage and mortality to the fish. The liberated miracidia swims about after it leaves the fish until it finds an appropriate snail host. It penetrates the snail, undergoes several changes which give rise to cercariae which leave the snail and swim about until they find an appropriate fish host. The cercariae penetrates the fish, makes its way to the heart and gill blood vessels where it developes into a mature worm and repeats the cycle of laying eggs. There are no known drugs or chemicals which can control this parasite in the fish. Preventing infections by eliminating snails and fish (which act as a reservoir of infection) in the water supply is the best control procedure known at this time.

Fungus

Dermocystidium salmonis

This fungus (some believe it is a protozoan) in California has been found in king salmon, steelhead and rainbow trout. On one occasion it was associated with a severe die-off in catchable size rainbow trout. The fungus forms pseudocysts in the gills and in the skin where it resembles "Ichs." Only a microscopic examination can reveal the true identity of this parasite. When the cyst-like structures mature, they appear to rupture, releasing the contents which are small spore-like cells. Presumably these cells are the infective agent. Many bacteria are found among these spore-like structures contained in the pseudocysts. These are believed to be secondary invaders and their role in the mortalities is not understood. No treatment for *Dermocystidium salmonis* has been reported.

Gas Bubble Disease

When gases from the air (oxygen, nitrogen, carbon dioxide, etc.) are under pressure, the amount dissolved in the water increases and it becomes supersaturated with the gases. Nitrogen, when present in a supersaturated state, enters the fishes body and if the fish moves to a position of less pressure, gas bubble disease occurs. Under such a condition, bubbles of gas (usually nitrogen) form and may be seen in various tissues such as the fins, skin, eyes, blood, etc.

In sac fry the gas causes coagulated yolk and subsequent death. Swimming "buttoned up" fry may develop a large gas bubble in the body cavity and then float or swim about erratically. Fingerlings, catchable and larger fish usually suffer eye damage with resulting blindness and even displacement and loss of the entire eye. Mortalities among this size fish are usually due to gas embolism in the circulatory system which blocks the blood from reaching vital areas of the body.

Gas bubble disease can sometimes be observed in fish being reared in spring or well water where the water has been under pressure and absorbed an excess of nitrogen and other gases. Occasionally air enters the intake of a water pump and is placed under pressure and again may cause gas bubble disease. There are also a number of other methods which may result in air supersaturating the water.

Supersaturated water may be returned to equilibrium by allowing it to pass through a gas exchanging or aerating tower, a device which allows the water to be broken up into many divided particles by falling through a series of screens mounted one above the other. There are other designs of gas exchangers that attempt to accomplish the same purpose: release the excess gas in the water. Occasionally in pumped water fish hatcheries, excess gas will dissipate enough not to cause any problems if the first pond in a series is kept empty and fish stocked in the remaining ponds in the series.

Virus Diseases

Virus diseases of fish have most likely been causing problems in fish hatcheries for a long time, but they have only recently been recognized after techniques were developed for isolating and identifying the viruses which cause them. To date three viral diseases of salmonids have been discovered which are of concern to fish culturists. All three are characterized by their ability to cause severe mortalities in young fish reared under hatchery conditions. There is concern that these diseases may have the same affects on wild populations of fish. However, at present nearly all the information available deals with virus diseases in the hatchery and one can only speculate about what is happening with wild populations.

The three diseases of concern are *Infectious Pancreatic Necrosis* (IPN), *Infectious Hematopoietic Necrosis* (IHN), and *Viral Hemorrhagic Septicema* (VHS).

Infectious Pancreatic Necrosis (IPN)

Causative agent: an RNA virus with a size of approximately 55 mμ which has been tentatively placed in the Reovirus group.

Recorded host species: rainbow trout, brook trout, cutthroat trout, coho salmon and Atlantic salmon.

Geographic range: IPN has been reported in trout cultural areas of the following countries: France, Italy, United Kingdom, Denmark, Sweden, Japan, Canada, and the United States.

Detection and Diagnosis: Symptoms observed in infected fish include exophthalmia, distension of the abdomen, darkening in color, and hemorrhages in the ventral surface, particularly at the bases of fins. *Internally,* the liver and spleen are pale and petechiae often are found in the anterior visceral mass. *Histologically,* the pancreas has been found to be the target organ showing pronounced necrosis. Some degenerative changes have also been observed in the hematopoietic tissue of the kidney.

Infectious Hematopoietic Necrosis (IHN)

Oregon sockeye disease (OSD) and Sacramento River chinook disease (SRCD) are considered the same as IHN since their causative agents cannot be distinguished from the virus causing IHN.

Causative agent: an RNA virus with a size of approximately 158 x 90 mμ which has been tentatively placed in the Rhabdovirus group.

Recorded host species: rainbow trout, steelhead trout, chinook salmon and sockeye salmon.

Geographic range: IHN has been found in some of the anadromous fish populations in California, Oregon, Washington, and British Columbia. IHN has also been isolated from a commercial hatchery complex in Eastern Washington. The few outbreaks of IHN in other parts of the United States have all been traced to the Eastern Washington source.

Detection and Diagnosis: Symptoms observed in infected fish include exophthalmia and distended abdomen, and hemorrhages in the musculature at the base of the pectoral fins. *Internally,* the liver, kidney and spleen are pale and the body cavity often fills with fluid. *Histologically,* the kidney is the primary target organ showing extensive degeneration and necrosis. Other organs showing involvement are the spleen and pancreas.

Viral Hemorrhagic Septicema (VHS)

Causative agent: RNA virus with a reported size of 70 x 180 mμ which has been placed in the Rhabdovirus group.

Recorded host species: rainbow trout, brown trout, brook trout.

Geographic range: the trout growing regions of Europe.

Detection and Diagnosis: symptoms observed in infected fish include exophthalmia, gill anemia, distended abdomen, and darkening in body color; hemorrhages may occur in the eyes, muscles and/or at the base of the pectoral fins. Histology has shown the kidney to be the primary target organ with pathology also observed in liver, spleen, skeletal muscle and adrenal cortex.

Identification and Control

Although these diseases are caused by three distinctly different viral agents, they do have some common characteristics. For example, all three can produce high mortalities in juvenile fish with survivors of an epidemic becoming carriers of the disease. These carrier fish are able to infect healthy fish which are placed in contact with them or in the water supply below them. The causative agents for these diseases can also be transmitted with spawning products. It is not clear at present whether this is a true ovarian transfer or mechanical transfer with the egg being contaminated during spawning or fertilization.

Another area of similarity is the external and internal symptoms observed in fish with viral diseases. The symptoms are so alike it is not possible to differentiate between the three diseases on the basis of symptoms alone. In fact the diagnosis of a viral disease should not be based solely on external and internal symptoms.

Diagnosis of viral diseases is a complicated procedure which involves the use of fish cell cultures. These cell cultures are grown in tubes and bottles in the presence of a medium containing all the essential elements for cell growth. For diagnosis, sample material from fish to be tested is placed in the presence of fish cell cultures. Viruses can only grow inside a living cell. Therefore, if a virus is present in the sample material, it multiplies in the cells causing changes which can be recognized (Figures 91 and 92). This is considered presumptive evidence for the presence of a virus. Further tests are carried out to confirm the presence of the virus and to determine its identity. The total sequence of events from sample taking to final tests may take as long as 2 to 3 weeks.

Most of the diseases found in fish can be controlled with some form of prescribed treatment. This is not the case with viral diseases. No drug or chemical has been found which has any effect on the course of a viral disease. The lack of success in treating these diseases has placed added emphasis on measures to prevent their introduction. In most cases introduction of viral diseases can be traced to the movement of diseased fish or eggs. To protect against this route of contamination, eggs should not be accepted until the broodstock has been checked and certified as virus free. The same rule should be applied to fish. They should not be accepted until they have been checked and found free of virus diseases.

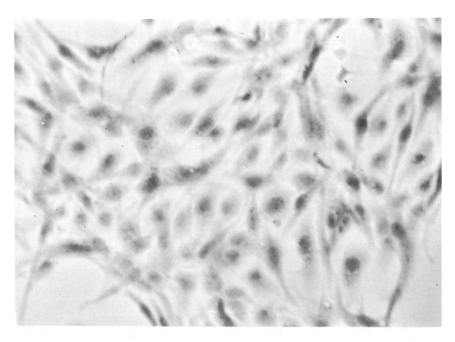

FIGURE 91—Monolayer culture of rainbow trout gonad (RTG-2) cells after 72 hours of incubation. *Photograph by William Wingfield, 1973.*

Close adherence to these two criteria will provide adequate protection for hatcheries with a closed water supply. Hatcheries on an open water source have the added concern that diseased fish may be planted into their water supply. There is no real safeguard against this situation other than regulations which prohibit the indiscriminate dispersal of fish with diseases. When fish in a hatchery are found to have a viral disease, the decision must be made whether or not to attempt disinfection of the installation. This decision will depend on the water supply of the hatchery and the source of fish and eggs. If the hatchery has an open water supply which contains infected fish, there is little sense in attempting disinfection until the water source is changed or until all fish have been removed. When fish or eggs for a hatchery (example: anadromous hatchery) are taken from a native stock of salmonids that have a viral disease, there is little use in disinfecting the hatchery since each new shipment of fish or eggs reinfects the installation.

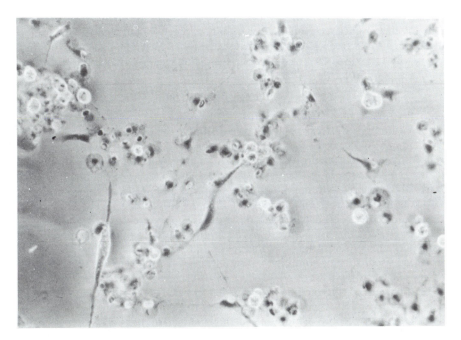

FIGURE 92—Monolayer culture of RTG-2 cells after 72 hours exposure to IPN virus. Cells are showing cytopathic changes which are characteristic of the IPN virus. *Photograph by William Wingfield, 1973.*

If the above factors are not a problem and disinfection of the hatchery is to be attempted, the following procedure has given best results:

1. Remove all fish from the hatchery before starting clean up procedure.
2. Thoroughly clean all ponds, troughs, and equipment.
3. After clean up is completed, disinfect the entire facility with 100 ppm chlorine.
4. Allow the hatchery to remain dry for at least three months after treatment.
5. After drying period, again disinfect entire installation with 100 ppm chlorine.
6. Finally, when restocking, be sure all fish and eggs brought in have previously been checked and found free of viruses.

To compromise on any one of the above steps greatly reduces the chances for a complete disinfection of the hatchery. Even careful adherence to the outlined procedure is no guarantee of total success. Experience has shown that once a virus disease becomes established in a hatchery it is very difficult to eradicate.

The difficulty of cleaning up contaminated installations plus the lack of effective treatments for viral diseases should make the fish culturist very cautious about any fish or eggs received at his installation. They should always be checked and certified virus free before they arrive at the hatchery. The general rule for dealing with viral diseases is best stated in the old adage: "An ounce of prevention is worth a pound of cure."

METHODS OF TREATMENT

Dip Method

In this method a strong solution is used for a relatively short period. Dip treatments are best carried out in wooden tubs or boxes. Metal tubs are often attacked by chemicals and products formed or liberated may be toxic to fish. A wooden tub of about 12 gallon capacity is recommended. Ten gallons of water is carefully measured into the tub and the required amount of chemical is added and mixed in. A net with enough bag to allow the fish to be entirely covered by the solution is used. The fish are netted, surplus water is allowed to drain off, and the net containing the fish is placed in the solution. The lip of the net should not be submerged, to prevent fish from escaping into the tub. When the time is up the net is raised, surplus solution is allowed to drain back into the tub, and the fish are placed in fresh water.

It is important that treated fish be returned to troughs or tanks that have been cleaned or preferably disinfected. Treated fish should never be returned to troughs, tanks, or ponds which derive their water from sources that contain diseased fish.

Bath Method (Prolonged Dip)

In this manner of treatment a weaker concentration is used for a longer period than in the dip method. It is often carried out in troughs and may be used in tanks, ponds, and flumes where the water may be turned off and the volume determined. When the desired volume is reached, the water is turned off and the system is checked for leaks. If the water level remains stable and the volume has been determined, the proper amount of chemical is added and mixed in. The length of treatment will vary with the water temperature, condition of the fish, and the chemical being used. In many cases it is necessary to aerate the water during treatment.

Aeration may be accomplished in several ways. The hand aerator is a primitive but satisfactory method for aerating troughs or small tanks. It is not suitable for ponds or flumes of large size or volume. Gasoline or electrically operated air compressors, fitted with rubber hoses and aerator stones, have been used successfully in troughs and small tanks.

One of the best methods of aerating a large pond or tank is to use a motor driven pump which will recirculate the water. The intake hose is placed at one end of the pond and the discharge hose at the other end. The discharge hose is directed at the side or end of the pond, thus breaking up the flow and aerating the water.

Flush Treatment

This method has become very popular, due to its simplicity. It may be used in raceways, tanks, and troughs. Without removing the fish, a measured amount of a chemical solution is added at the inlet and allowed to flush through the pond, tank, or trough. The amount of chemical, rate of flow, and depth of water must be established on an individual hatchery basis by trial before large-scale flush treatment methods can be employed.

A stock solution of copper sulfate, varying from ½ to 1½ pounds per gallon of water, has been flushed through standard raceway ponds with a flow of 1 to 3 cfs and found to control bacterial gill disease. "Gyros" have been controlled by flushing ½ to 1½ gallons of acetic acid through ponds with a flow of 1 to 3 cfs. Malachite green at a rate of 3–4 ounces of dry ingredient dissolved in 5–10 gallons of water for a flow of 2–3 cfs has aided in combatting a variety of external organisms from "Ichs" and *Epistylis* to fungus and columnaris.

It has been found that lowering the water level in raceways evens out the flow and eliminates dead water spots, thus insuring a better distribution of the treating agent.

Disinfection

Occasionally the need for disinfection of equipment arises. One of the best and cheapest disinfectants is chlorine. A solution of 200 ppm will sterilize equipment in 60 minutes. A solution of 100 ppm will require several hours for complete sterilization. Chlorine concentration is reduced by the presence of organic material such as mud, slime, and plant material. Therefore, for full effectiveness, it is necessary to clean thoroughly equipment to be exposed to the chlorine solution. A chlorine solution will lose its strength by exposure, so it is necessary to add more chlorine or make up fresh solutions. Containers used in sterilizing small pieces of equipment should be painted with asphaltum before being used.

Chlorine is toxic to trout. If troughs, tanks, or ponds are disinfected, the chlorine must be neutralized before it is allowed to pass undiluted through waters containing concentrated numbers of fish.

One gallon of 200 ppm chlorine solution may be neutralized by adding 5.6 grams of sodium thiosulfate. Neutralization may be determined by the use of starch-iodide chlorine test papers (obtainable from LaMotte Chemical Products Company) or by the use of orthotolidine solution. A few drops of orthotolidine are added to a sample of the solution to be tested. If the sample turns a brown color, chlorine is present. Absence of color means that the chlorine has been neutralized.

Chlorine may be obtained as sodium hypochlorite in either liquid or powdered form. The latter is the more stable of the two, though more expensive.

The amount of chlorine to be added to water in making up a predetermined solution is based on the percentage of available chlorine in the product used. As an example, HTH powder may contain either 15, 50, or 65% available chlorine. It would, therefore, require the following amounts to make a 200 ppm solution.

Two ounces of 15% available chlorine HTH powder to 10½ gallons of water.

One ounce of 50% available chlorine HTH powder to 18 gallons of water.

One ounce of 65% available chlorine HTH powder to 23¼ gallons of water.

Drug and Feed Mixing

Treatment of some diseases, such as columnaris, ulcer disease, and furunculosis, requires the feeding of drugs. This is accomplished by mixing the drug with the feed. The amount of drug to be fed is relatively small and thorough mixing is necessary to insure proper distribution. Fish should be hungry before they are administered drugs; it may be necessary to eliminate a feeding to insure that the drugged food is taken readily.

With the development of dry feeds it has been possible to buy feed containing the drug of choice. In most cases it is necessary to have the food containing the drug custom-milled. Fish of different sizes require varying amounts of feed and drug, and custom milling is necessary in order to get the proper dosage.

For example, 100 pounds of trout might require food amounting to 5% of their body weight. Thus, 5 pounds of feed would contain 10 grams of drug. Another 100 pounds of different sized trout might require 3% of their body weight of feed. Thus, for the same dosage they should get 10 grams of drug in 3 pounds of feed.

CHEMICALS AND THEIR USES
Salt

Salt is one of the oldest and most commonly used chemicals in the treatment of diseases of trout. Its therapeutic value is based on two of its actions. (1) It causes the mucus covering the fish's skin to be sloughed off, thereby removing organisms which are loosely attached to the mucus. (2) It raises the specific gravity of the water and changes the osmotic pressure which, in turn, causes many external parasites to burst.

Salt seems to exert the best action when used as a 3% solution. The length of the bath varies with the water temperature and condition of the fish. *Costia, Ichthyophthirius, Chilodon, Trichodina,* and *Epistylis* are all external protozoans which may be controlled in varying degrees by salt treatments. Usually two or three treatments are necessary.

Copper Sulfate

Copper sulfate, or bluestone, is another old standby. Properly used, it provides a most effective treatment. It is often used as a dip at a 1:2,000 concentration for 1 to 2 minutes. It may be necessary to add a small amount of acetic acid (¾ ounce) to a 10 gallon tub of the solution in hard waters. The acid prevents the copper sulfate from going out of solution and forming a white precipitate at the bottom of the tub. When the copper sulfate precipitates, part of its beneficial action is lost.

Copper sulfate is useful in controlling external bacteria causing fin rot, tail rot, and gill disease, provided the fish are treated before the bacteria have invaded the deeper tissues. Some external protozoans may also be controlled by copper sulfate; however, other treatments are more satisfactory in controlling these organisms.

Acetic Acid

This acid will eradicate many external parasites in a single treatment when used as a dip of 1:500 for 1 minute. Heavy infestations may require follow-up treatment.

External protozoans and the worm *Gyrodactylus* are effectively controlled by this treatment. Bacterial infections are not particularly susceptible to acetic acid. Each tub of solution should be aerated at intervals and the solution renewed after five or six batches (50 to 70 ounces per batch) of fish have been dipped. After the third or fourth bath, the immersion time may be increased to $1\frac{1}{2}$ minutes to compensate for dilution.

Formaldehyde (Formalin)

This chemical is used as a prolonged dip ($\frac{1}{2}$ to 1 hour) in troughs, tanks, and raceway ponds when the water volume can be determined with reasonable accuracy. Because of the duration of the treatment, it may be necessary to aerate the water while the fish are being treated. A concentration of 1:4,000 is best, although a dilution of 1:6,000 may sometimes be satisfactory.

Formaldehyde is most useful in treating for external protozoans. One advantage in using formaldehyde lies in not having to handle the fish; however, in certain types of water it may cause above normal treatment losses. A word of caution to those who have not used formaldehyde is to try it in a single trough a day before treating an entire hatchery, to determine if a significant loss will follow.

Sulfamerazine

Over the years different sulfa drugs have been used in treating bacterial diseases (furunculosis, red-mouth, etc.) among trout and salmon. Presently, sulfamerazine and sulfamethazine are the most widely used and appear to have been used virtually interchangeably. In California, sulfamerazine has been the most widely used and at a level which provides 10 grams of drug/100 pounds of fish per day (in a dry diet) for 5 to 10 days. Moist pellet diets, it has been suggested, should contain 5 grams/100 pounds of fish per day for 10 days. Sulfa drugs have occasionally been reported to be toxic to silver salmon so when used with this species one should watch for increased mortalities and discontinue its use if encountered.

Terramycin

Terramycin (a trade name for Oxytetracycline) is an antibiotic that has come into wide and effective use against some internal and external bacterial diseases of trout and salmon. When used against internal diseases such as furunculosis, it is milled into a dry or moist commercially prepared diet at a rate which will allow 4 grams of pure antibiotic to be fed to every 100 pounds of fish per day. The standard treatment lasts for 10 days. TM-50 or TM-50D (water dispersible) is the form usually added to commercially prepared fish food. Each pound of TM-50 (or TM-50D) contains 50 grams of pure antibiotic.

Occasionally it may be necessary to add Terramycin to small amounts of food. This may be done by mixing an appropriate amount of TM-50 or TM-50D or Terramycin Soluble Powder in a gelatin solution (1 ounce gelatin to 1 quart of warm water) and spraying it over the daily food ration. The water soluble powder concentrate of Terramycin is the easiest to work with. This form may be purchased in 4 ounce preweighed packages each of which contains 25.6 grams of antibiotic. As much as two packages of this form may be dissolved in 1 quart of hot gelatin solution. The treatment of external bacterial infections such as columnaris and gill disease are often successfully treated in troughs and tanks with 30 minutes to 1 hour exposures to the Terramycin Soluble Powder Concentrate in solution. One successful concentration uses 1.75 grams of concentrate (as it comes from the package) per 10 gallons of water. In tanks and troughs, the technique requires lowering the water below the required volume, adding the Terramycin (dissolved in some water), then allowing the water to fill to the desired level, and then turned off. After the proper length of time, the water is turned on and allowed to flush the tank or trough with the normal water flow.

Furox 50

Furox 50 (formerly known as NF 180) is another drug which has been used with some success against internal bacterial infections such as furunculosis. It contains 11% furazolidone and is fed at a rate of 25–50 grams (as it comes from the bag) per 100 pounds of fish. Occasionally toxic effects have been noticed (a rise in losses) with this drug so the fish should be carefully observed during its use. Furox 50 is used by mixing it with corn oil heated to 100° F. at a rate of ½ to 1 pound of Furox 50/1000 pounds of fish with the drug being mixed in 750 cc of corn oil for every 50 pounds of food to be fed.

Sulfomerthiolate

This organic mercurial compound has been used successfully in experimental work for the disinfection of trout eggs during all stages of their development up through the eyed stage. Sulfomerthiolate is especially used in disinfecting eggs that have been exposed to furunculosis. The concentration recommended is 1:5,-000. It is best to disinfect eggs either soon after they have become water hardened or after they are eyed, although eggs may be treated during the tender stage if handled carefully. As many as 50,000 eggs may be treated in 3 gallons of the solution. Two methods used for treating eggs are the following:

1. Dip eggs for 10 minutes in a 1:5,000 solution. The eggs should be moved about every 2 minutes to insure complete contact with the solution. A wooden, enamel, or asphalt-painted container should be used to hold the solution. Several baskets of eggs may be treated with the same solution if it is well aerated and not more than a few hours old.

2. An entire trough of egg baskets may be treated at one time with the bath method. It is necessary to calculate the amount of water in the trough and to add the amount of sulfomerthiolate required to produce a 1:5,000 concentration. The bath is continued for 10 minutes. During the treatment it is necessary to aerate or recirculate the water. Both the sulfomerthiolate *powder* and *solution* are quickly destroyed by direct sunlight and must be kept shaded at all times.

Acriflavine

If sulfomerthiolate is not available, acriflavine may be used for disinfecting eggs. A concentration of 1:2,000 is recommended. The eggs are held in the treating solution for 20 to 30 minutes. It is also possible to disinfect the eggs with a 10 minute treatment in a 0.185% solution.

Wescodyne

Wescodyne is a complexed iodine-detergent formula manufactured by the West Chemical Company. It is reputed to be active against bacteria, viruses, molds, yeasts, and fungi.

A 1:300 (4.5 fluid ounces as it comes from the bottle per 10 gallons of water) concentration for 10 minutes appears to be safe for use with trout and salmon eggs. After the eggs have been so treated, they may be put away in baskets or incubators and the water allowed to flow over them. If the eggs are to be shipped, they should be rinsed in flowing water before packing. Either green or eyed eggs may be treated. Not more than 25 ounces of eggs per gallon of solution should be treated at a time. The solution which is an amber color may be used until it approaches a light yellow when it should be discarded and a fresh solution made.

In soft water, 0.5 grams of sodium bicarbonate (baking soda) should be added to each gallon of solution. This will keep the solution from dropping to an acid state. If there is any question of the nature of the water, add the sodium bicarbonate. In some areas Wescodyne is used on eggs at a concentration of 1:150 though its need is not clear. Wescodyne at a concentration of 1:20,000 is harmful to fish. Dispose of solutions in a safe place.

Malachite Green

This dye is well known to the science of bacteriology for its bactericidal properties. In 1936, it was successfully used in controlling fungus on trout. It was found that a 1:15,000 solution of malachite green used as a 10 to 60 second dip for two or three treatments cleared up fungus infections. Fungus on trout is usually considered a secondary invader and unless the primary cause is controlled, redevelopment of the fungus may occur. It is important that the malachite green specified be of low zinc content.

Malachite green is also used to control fungus on salmonid eggs. Three methods of treating eggs have been developed.

1. The flush method involves the use of a stock solution (1½ ounces malachite green dissolved in 1 gallon of water). Three ounces of this solution are poured into the head end of the trough. Water inflow is set at 6 gallons per minute. This process is repeated every other day, although it may be repeated daily if required.

2. The dip method involves the use of a solution of known concentrations and requires the egg baskets to be drained and placed in the dip for varying lengths of time. This method is not used often, since the handling may cause shock to the eggs, with subsequent losses.

3. The prolonged dip method is probably the best, but requires a certain amount of work in setting up and calibrating the equipment to be used. However, any hatchery troubled with a severe fungus condition on eggs may be better off to use this method rather than the others previously mentioned.

This method requires the use of a constant flow siphon which adds malachite green to the water at a 1:200,000 concentration for 1 hour. Complete details for using this method are given by Burrows (1949).

TABLE 18

Solution Chart for a Standard California Hatchery Trough

Depth in inches	Gallons	Cubic feet	Cubic inches	Cubic centimeters	Acetic acid 1:500 1- or 2-minute dip		Copper sulfate 1:2,000 1- or 2-minute dip			Formaldehyde 1:4,000 ½ to 1 hour bath	
					Ounces	cc	Ounces*	Grams*	cc.†	Ounces	cc
1	13.29	1.777	3,072	50,340	3.40	100.6	0.88	25.17	211.0	0.43	12.5
2	26.59	3.554	6,144	100,681	6.80	201.2	1.76	50.34	422.0	0.86	25.17
3	39.87	5.331	9,216	151,022	10.20	301.8	2.64	75.52	633.0	1.30	37.75
4	53.16	7.108	12,288	201,363	13.60	402.4	3.52	100.69	844.0	1.73	50.34
5	66.45	8.885	15,360	251,704	17.00	503.0	4.40	125.87	1,055.0	2.17	62.9
6	79.74	10.662	18,432	302,045	20.40	603.6	5.28	151.04	1,226.0	2.60	75.5
7	93.03	12.439	21,504	352,386	23.80	704.2	6.16	176.21	1,477.0	3.04	88.0
8	106.32	14.216	24,576	402,726	27.20	804.8	7.04	201.39	1,688.0	3.47	100.68

* Dry weight.
† Solution of one pound of copper sulfate dissolved in one gallon of water.
‡ 10 percent PMA solution (10 grams PMA dissolved in 100 cc. of water).

ANESTHETICS AND THEIR USE

Anesthetics have been used for some time by fisheries workers as an aid in fin clipping, spawn taking, and fish transportation. The degree of anesthesia required varies with the use. Transportation requires a light anesthesia, whereas a deep anesthesia is best for spawn taking.

A variety of drugs and chemicals have been used. These include cresol, ether, carbon dioxide, chlorotone (Chlorobutanol), sodium amytal, methyl pentynol, quinaldine, MS-222 (tricaine methane sulfonate), and others.

Presently, only MS-222 and carbon dioxide are cleared by the Food and Drug Administration for use on food fishes.

MS-222 (Tricaine Methane Sulfonate)

This odorless, water-soluble drug comes in powdered form. It is nonprescriptive and has a good storage life unless placed in solution. No vehicle is required for making an aqueous solution. The measured amount of powder is simply sprinkled into the water and stirred. It has a safe exposure time in that fish may be left in the solution for a considerable duration. Recovery when placed in fresh water is quite rapid.

MS-222 is restricted to a level of 15 to 330 ppm and the fish must not be used within 21 days following the anesthesia for food purposes.

Carbon Dioxide

Carbon dioxide was originally used and was made by the reaction of sodium bicarbonate and sulfuric acid. Most users presently buy cylinders of carbon dioxide and add it through a carborumdum stone or some other device which adds the compound to water in a finely divided state. The rate at which it is added is usually determined on a trial basis and will reflect the size of the fish being handled and the water temperature.

TABLE 19

Dip Treatments

Acetic acid 1:500			Copper sulfate 1:2,000					
Ounces	Cubic centimeters	Gallons of water	Grams*	Ounces*	Ounces†	Cubic centimeters†	Gallons of water
0.27	7.0	1	1.89	0.067	0.53	15.88	1
1.35	37.8	5	9.46	0.33	2.68	79.40	5
2.70	75.7	10	18.92	0.67	5.36	158.80	10
4.05	113.5	15	28.38	1.01	8.05	238.20	15
5.40	151.4	20	37.84	1.35	10.73	317.60	20

* Dry weight.
† Solution of one pound of copper sulfate dissolved in one gallon of water.

NETS AND SEINES

The capture of fish with nets or seines has been practiced by fishermen for centuries. Early fish nets were woven of raffia, hemp, or reed grasses. Cotton nets have been used for many years. At present, however, nets woven of nylon twine are in greatest use. There are many known methods of weaving or mending nets throughout the world, but all are basically similar. Most nets are made by machine but their mending must be done by hand; this will probably always remain a part of the fisherman's work.

TABLE 20
Salt Solution Chart for a Standard California Trough

Depth in inches*	Percentage	Salt (pounds)	Salt (grams)
3	1	3.30	1,510
3	2	6.60	3,020
3	3	9.90	4,530
4	1	4.41	2,013
4	2	8.82	4,027
4	3	13.23	6,040
5	1	5.54	2,517
5	2	11.09	5,034
5	3	16.64	7,551
6	1	6.65	3,020
6	2	13.31	6,040
6	3	19.97	9,061

* Depth measured at center of trough.

Net mending is an age old art and proficiency in this field is attained only after considerable instruction and practice. Fish culturists are not expected to be highly skilled in the art of net mending. Minor repairing of nets used at hatcheries, though, is part of the hatcheryman's duties and some knowledge of net mending will be useful. A recommended reference and guide to net mending can be obtained by writing to the U. S. Fish and Wildlife Service, Washington, D. C. The publication is "Methods of Net Mending", Fishery Leaflet No. 241 (Knake, 1947).

TABLE 21
Twisted Ropes Break Test Chart

Diameter	Nylon Approx. lb.	Polyethylene Approx. lb.	Polypropylene Approx. lb.	Manilla Approx. lb.
3/16"	1,050	760	800	450
1/4"	1,800	1,250	1,300	600
5/16"	2,750	1,800	1,900	1,000
3/8"	4,000	2,550	2,700	1,350
7/16"	5,500	3,350	3,500	1,750
1/2"	7,100	4,200	4,400	2,650
9/16"	9,075	5,200	5,400	3,450
5/8"	10,900	6,200	6,500	4,400
3/4"	15,600	8,600	9,000	5,400
7/8"	21,000	11,500	12,000	7,700
1"	27,000	14,300	15,000	9,000

Braided Ropes Break Test Chart

Diameter	Nylon Approx. lb.	Polyethylene Approx. lb.	Polypropylene Approx. lb.
3/64"	90		
1/16"	120		
5/64"	155		
3/32"	275	215	225
7/64"	357		
1/8"	479	260	275
9/64"	556		
5/32"	664	440	465
3/16"	1,119	700	735
1/4"	1,665	1,000	1,060
5/16"	3,000	1,600	1,680
3/8"	3,300	2,000	2,100
1/2"	4,200	3,500	3,675

Net Webbing

Seines have been made by hanging the net webbing on manila hemp or cotton ropes. The ropes now generally used are nylon, polyethylene, or polypropylene. These ropes are much stronger, rot proof, abrasion resistant, and are not damaged by oil, gasoline, or most chemicals. They can be obtained either braided or twisted. The braided rope has the advantage that it will not kink or untwist. The approximate pound breaking test of various size ropes indicates that nylon is the stronger rope (Table 21). The upper or float line on a seine secures the floats which cause the line to float and the net to hang in a vertical position in the water. The lower or lead line is equipped with weights which sink the lower portion of the net. Net floats are made of cedar, cork, glass, plastic, or aluminum. Cedar or cork floats are in common use. Weights for the lead line are nearly always made of lead. The action of a net in the water can be regulated, to some extent, by the size and number of floats or weights used. Nets which are drifted in the water have a greater number of floats than those which are still-fished or are drawn through the water.

Nets are woven of seine twine, which comes in a variety of sizes (Figure 93A).

Nets are measured by the mesh, either stretched mesh or square mesh (Figure 93B). Considerable latitude is allowed in the size of twine used. The smaller the mesh, the finer the twine.

When ordering nets, it is always necessary that the following information be given: mesh, square or stretched (Figure 93B); twine, size (Figure 93A); twine, soft-, medium-, or hard-laid; length in meshes, depth in meshes; selvage, single or double (Figure 93C); floats, size and spacing; weights, size and spacing; lines, size rope for both float and lead lines.

Nylon netting is in common use because it is much stronger and more durable than cotton (Table 22). The nylon is much superior so a smaller size twine can be used in seine material which will make the seine easier to pull through the water.

TABLE 22
Seine Twine Break Test Chart

Size	Nylon Approx. lb.	Cotton Approx. lb.
3	22	--
4	32	--
5	45	--
6	52	12
7	59	--
9	85	18
12	92	24
15	125	31
18	158	36
21	175	41
24	200	48
30	240	60
36	275	72
42	370	85
48	395	96
54	440	112
60	501	120
72	552	155
84	600	184
96	652	200
120	710	236

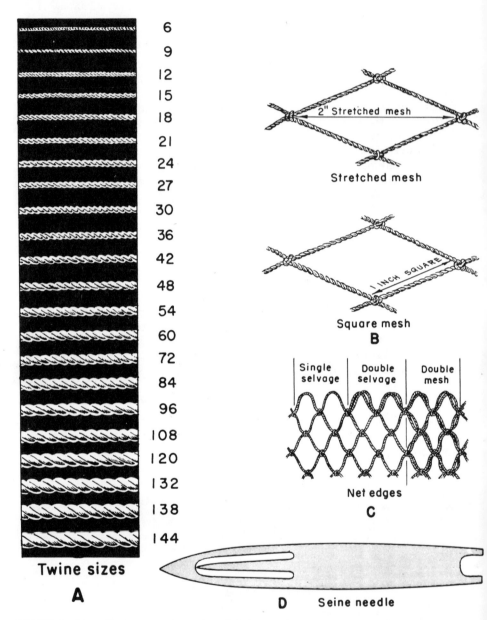

FIGURE 93—Net and twine nomenclature (actual size).

In nearly every instance the seine webbing will wear out much before the float and lead lines; even after the lines become unusable, the floats and leads remain in good condition. For this reason some hatchery managers obtain seine webbing of a size to meet their requirements and make up or hang their own seines, reusing the old rope, floats and leads. A mistake often made in ordering seines or making

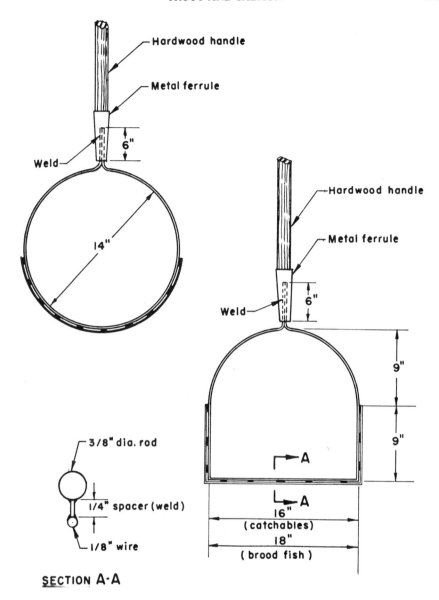

Hardwood handle

Metal ferrule

6"

Weld

14"

Hardwood handle

Metal ferrule

6"

Weld

9"

9"

A

A

16"
(catchables)

18"
(brood fish)

3/8" dia. rod

1/4" spacer (weld)

1/8" wire

SECTION A-A

FIGURE 94—Dip net frames.

them is specifying a seine that is too short and not deep enough to allow it to bag properly when used in fishing.

Seines of many sizes and mesh are used at hatcheries depending upon the varied facilities and operations. The seines most commonly used are approximately 30 feet long and 6 feet deep made of ½ inch and 1 inch stretched mesh, nylon No. 5 twine.

The concrete raceway type rearing ponds in new hatcheries, and the replacement of old type earthen ponds in California hatcheries will make the use of seines obsolete. The fish will be managed with crowders (hand or mechanical) and fish pumps. The net principally used at the hatcheries will be the dip net.

Nets, like all gear, will wear out and some losses cannot be avoided. Proper care and treatment, however, will greatly prolong the life of a net. It should be washed, spread out, and dried in the shade each time after use, then rolled up and stored where the sun will not shine on it. The agent most destructive to nets kept in water for considerable periods is a microscopic form of life that digests (eats) cotton or similar material. The agent is most active when a damp net is piled in a heap during warm weather.

Remember, nets should always be washed, spread out, and dried immediately after use. It is well recognized that the life of a net can be prolonged greatly if it receives proper care.

Dip Nets

Dip nets are usually made to meet individual requirements and consist of a metal frame on which the dip net bag is hung and a hard wooden handle. Insofar as it is known, dip net frames suitable for hatchery use are not commercially manufactured and usually the frames are fabricated of metal rod or heavy wire and attached to a handle suitable for the job. In selecting dip net handles, it is well to consult a wholesale hardware catalog in which a variety of suitable wooden handles with metal end ferrules can be found (Figure 94).

Many sizes of dip nets are used in hatcheries and with planting equipment. The size dip net most commonly used is approximately 14 inches x 14 inches by 12 inches deep. Some hatchery managers prefer a dip net that is about 8 inches x 12 inches for dipping fish from a planting tank and pouring them into a planting bucket to carry to a stream, while others prefer the larger dip nets. Larger dip nets are used for brood fish (similar to examples shown in Figure 94).

Dip nets are essentially seine material and require the same treatment and care as seines. They are generally woven in rectangular shape to fit the frames. For use with catchable trout they are usually made with ½ inch or 1 inch stretch mesh nylon. For brood fish they could be 2 inches stretched mesh nylon.

Catalogs for seines, netting, dip net bags, and allied materials are available from the following vendors to aid in ordering equipment:

Memphis Net and Twine Co.
2481 Mathews Ave., Memphis, Tenn. 38108

West Coast Netting, Inc.
14929 Clark Street, City of Industry, CA 91745

Nylon Net Company
P. O. Box 592, Memphis, Tenn. 38101

Sterling Marine Products
7 Oak Place, Montclair, New Jersey 07042

Consolidated Net and Twine Co.
P. O. Box 223
University Station—Seattle, Washington 98105

Marion Textiles, Inc.
48 West 37th Street—New York, N. Y. 10018

Nichols Net and Twine Co.
Rural Route #3, Bend Road
East St. Louis, Illinois 62201

THE CLASSIFICATION OF FISHES

All known living things have been given scientific names and classified by categories to show their evolutionary relationships.

The fisheries manager will do well to master an understanding of the classification of fishes generally, and particularly to know the relationships and names of the fishes with which he deals. Such an understanding will help him in his discussions with the interested public, and will enable him to more competently handle various problems that may arise.

The basic categories used to classify all animals (and plants), listed in order from the smallest unit to the largest grouping, are as follows: species, genus, family, order, class, and phylum.

A species may be defined as a group of interbreeding individuals not ordinarily interbreeding with another such group; a systematic unit including geographic races and varieties, and included in a genus. A species may include two or more subspecies.

The scientific name of every species or "kind" of animal consists of two parts, first the name of the genus and second the name of the species. Thus, the scientific name of the rainbow trout is *Salmo gairdnerii,* and it is a member of the Family Salmonidae, which includes all of the trouts, salmons, and chars. The scientific name is always written in italics. The name of the individual who described the species follows the scientific name (without italics), but is often omitted in nontechnical publications.

Ichthyologists are by no means agreed upon the proper arrangement of the various categories of classification for fishes. As new facts, which lead to new concepts of relationships, are uncovered, changes are made in names and in the grouping of the various categories of classification.

In accordance with general custom, in keys and lists the most ancient, simplest, or most primitive types are listed first, followed by a sequence ending with the most highly specialized, recent forms.

In addition to their scientific names, most game fishes and some nongame ones have been given common names. Scientific names are given and recognized only according to established rules of nomenclature, but anyone is "free" to call a fish by any common name he chooses. Thus, although the scientific names of fishes have varied to some extent as new facts have been uncovered, common names have varied even more widely, depending upon local usage, appearance based on habitat, sexual development, or sexual differences, superficial or supposed resemblance to other forms, or merely the whim of the individual. For example, other common names that have been used for the king salmon are black salmon (applied to individuals that have become dark because of long presence in fresh water), chub salmon (applied to young males), dog salmon or hookbill (applied to males with hooked snouts), silver salmon (applied to young fish fresh from the ocean in the Sacramento River system), chinook salmon, spring salmon, quinnat salmon, and tyee salmon.

In recent years serious attempts have been made by state authorities and national groups, notably the American Fisheries Society, the Society of Ichthyologists and Herpetologists, and the Outdoor Writers Association of America, to establish "official" common names for at least the principal food and game fishes of the United States. The most recent such list for California freshwater and anadromous fishes is that by Shapovalov, Dill, and Cordone (1959).

"Keys" to all the freshwater and anadromous fishes, which will enable the individual to identify a specimen in hand in accordance with differentiating characters, are now also available (Kimsey and Fisk, 1960). Those interested particularly in the relationships and characteristics of the trouts and salmons of California are referred to the publications by Shapovalov (1947) and Wales (1957).

A current list of the common and scientific names of the salmons and trouts of California follows. Forms which have been introduced into California waters are denoted by an asterisk (*), and those which are not currently being propagated in the State's hatcheries by a dagger (†).

1. Pink salmon † ....*Oncorhynchus gorbuscha*
2. Chum salmon † ....*Oncorhynchus keta*
3. Silver salmon..*Oncorhynchus kisutch*
4. King salmon ....*Oncorhynchus tshawytscha*
5. Sockeye salmon (anadromous form †);
   Kokanee salmon (freshwater form *)............*Oncorhynchus nerka*
6. Brown trout * ....*Salmo trutta*
7. Coast Cutthroat trout ....*Salmo clarkii clarkii*
8. Lahontan cutthroat trout ....*Salmo clarkii henshawi*
9. Paiute cutthroat trout † ....*Salmo clarkii seleniris*
10. Steelhead rainbow trout..............................*Salmo gairdnerii gairdnerii*
11. Kamloops rainbow trout * ....*Salmo gairdnerii kamloops*
12. Shasta rainbow trout ....*Salmo gairdnerii stonei*
13. Kern River rainbow trout † ....*Salmo gairdnerii gilberti*
14. Eagle Lake rainbow trout ....*Salmo garidnerii aquilarum*
15. South Fork of Kern golden trout................*Salmo aguabonita aguabonita*
16. Little Kern golden trout †...........................*Salmo aguabonita whitei*
17. Eastern brook trout *..................................*Salvelinus fontinalis*
18. Dolly Varden trout † ....*Salvelinus malma*
19. Lake trout * † ....*Salvelinus namaycush namaycush*

REFERENCES

Affleck, R. J. 1952. Zinc poisoning in a trout hatchery. Aust. Jour. of Marine and Freshwater Res., 3(2):142–169.

Allison, Leonard N. 1958. Multiple sulfa therapy of kidney disease among brook trout. Prog. Fish-Cult., 20(2):66–68.

American Public Health Association. 1955. Standard methods for the examination of water, sewage, and industrial wastes. Amer. Publ. Health Assoc., Inc., 10th ed., New York, 522 p.

Anonymous. 1949. Frontiers in nutrition. Chicago, Illinois, Dawe's Manufacturing Co. (and Affiliated Companies), October 96 p.

————. 1958. The anesthetic of choice in work with cold-blooded animals. Sandoz Pharmaceuticals, Hanover, New Jersey. 8 p.

Brockway, Donald R. 1950. Metabolic products and their effects. Prog. Fish-Cult., 12(3):127–129.

Burrows, Roger E. 1949. Prophylactic treatment for control of fungus *(Saprolegnia parasitica)* on salmon eggs. Prog. Fish-Cult., 11(2):97–103.

————. 1951. A method for enumeration of salmon and trout eggs by displacement. Prog. Fish-Cult., 13(1):25–30.

————. 1955. A vertical egg and fry incubator. Prog. Fish-Cult., 17(4):147–155.

Burrows, Roger E., Leslie A. Robinson, and David D. Palmer. 1951. Tests of hatchery foods for blueback salmon 1944–48. U.S. Fish and Wildl. Serv., Spec. Sci. Rept. 59, 39 p.

Burrows, Roger E., and Bobby D. Combs. 1968. Controlled environments for salmon propagation. Prog. Fish-Cult., 30(3):123–136.

Calhoun, Alex. 1953. Hypnotic drugs as an aid in fish transportation. Calif. Dept. Fish and Game, 3 p. (Mimeo.)

Carl, G. Clifford. 1941. Beware of the broken egg! A possible cause of heavy losses of salmon eggs. Prog. Fish-Cult., (53):30–31.

Davis, H. S. 1946. Care and diseases of trout. U.S. Fish and Wildl. Serv., Res. Rept., (12):1–98.

————. 1956. Culture and diseases of game fishes. Univ. of Calif. Press, Berkeley 332 p.

Earp, J. J., C. H. Ellis, and E. J. Ordal. 1953. Kidney disease in young salmon. Wash. Dept. Fish., Spec. Rept., Ser., (1):1–74.

Ellis, M. M., B. A. Westfall, and Marion D. Ellis. 1946. Determination of water quality. U.S. Fish and Wildl. Serv., Res. Rept., (9):1–122.

Embody, G. C. 1934. Relation of temperature to the incubation periods of eggs of four species of trout. Amer. Fish. Soc., Trans., 64:281–292.

Fish, Frederic F. 1934. Ulcer disease of trout. Amer. Fish. Soc., Trans., 64:252–258.

————. 1935. A western type of bacterial gill disease. Amer. Fish. Soc., Trans., 65:85–87.

Fish, Frederic F., and Robert R. Rucker. 1943. Columnaris as a disease of cold-water fishes. Amer. Fish. Soc., Trans., 73:32–36.

Foster, Fred J., and Lowell Woodbury. 1936. The use of malachite green as a fish fungicide and antiseptic. U.S. Bur. Fish., Prog. Fish-Cult., Memo I-131(18):7–9.

Gall, Grahm A. E. 1972. Rainbow trout broadstock selection program with computerized scoring. Calif. Dept. Fish and Game, Inland Fish. Adm. Rept., (72-9):1–22.

Greenberg, David M. 1957. Something of the anatomy and physiology of a trout. U.S. Trout News, Nov.–Dec.,:6–9.

Griffin, Philip J., S. F. Snieszko, and S. B. Friddle. 1952. A more comprehensive description of *Bacterium salmonicida*. Amer. Fish Soc., Trans., 82:129–138.

Halver, John E. 1957. Nutrition of salmonoid fishes. III. Water-soluble vitamin requirements of chinook salmon. Jour. Nutrition, 62(2):225–243.

Halver, John E., Donald C. DeLong, and Edwin T. Mertz. 1957. Nutrition of salmonoid fishes. V. Classification of essential amino acids for chinook salmon. Jour. Nutrition, 63(1):95–105.

Haskell, David C., and Richard O. Davies. 1958. Carbon dioxide as a limiting factor in trout transportation. New York Fish and Game Jour., 5(2):175–183.

Hayes, Frederick Ronald. 1949. The growth, general chemistry, and temperature relations of salmonid eggs. Quart. Rev. Biol, 24(4):281–308.

International Pacific Salmon Fisheries Commission. 1957. Annual report 1956. New Westminster, B.C., Canada, 32 p.

Johnson, Harlan E. 1951. Sulfamerazine in the control of columnaris in steelhead trout *(Salmo gairdnerii)*. Prog. Fish-Cult., 13(2):91–93.

Johnson, Harlan E., and Richard F. Brice. 1953. Effects of transportation of green eggs, and of water temperature during incubation, on the mortality of chinook salmon. Prog. Fish-Cult., 15(3):104–108.

Johnson, Harlan E., Clyde D. Adams, and Robert J. McElrath. 1955. A new method of treating salmon eggs and fry with malachite green. Prog. Fish-Cult., 17(2):76–78.

Keil, W. M. 1936. Selecting and developing a broodstock of trout. Prog. Fish-Cult., (25):1–4.

Kimsey, J. B., and Leonard O. Fisk. 1960. Keys to the freshwater and anadromous fishes of California. Calif. Fish and Game, 46(4):453–479.

Klak, George E. 1940. The condition of brook trout and rainbow trout from four eastern streams. Amer. Fish. Soc., Trans., (70):282–289.

Knake, Boris O. 1947. Methods of net mending. U.S. Fish and Wildl. Serv., Fishery Leaflet, (241):1–17.

Lagler, Karl F. 1952. Freshwater fishery biology. Wm. C. Brown Co., Dubuque, Iowa. 360 p.

Leach, Glen C. 1939. Artificial propagation of brook trout and rainbow trout, with notes on three other species. U.S. Bur. Fish., Fish. Doc. 955, 74 p.

Leitritz, Earl. 1959. Trout and Salmon Culture (Hatchery Methods). Calif. Dept. Fish and Game, Fish Bull., (107):1–169.

Lewis, R. C. 1944. Selective breeding of rainbow trout at Hot Creek Hatchery. Calif. Fish and Game, 30(2):95–97.

McCraw, Bruce M. 1952. Furunculosis of fish. U.S. Fish and Wildl. Serv., Spec. Sci. Rept., Fish., (84):1–87.

M'Gonigle, R. H. 1940. Acute catarrhal enteritis of salmonid fingerlings. Amer. Fish. Soc., Trans., 70:297–303.

McMullen, Richard J. 1948. Power egg picker. Prog. Fish-Cult., 10(1):30–31.

Nordqvist, Oscar. 1893. Some notes about American fish-culture. U.S. Fish Comm., Bull., 13:197–200.

O'Donnell, D. John. 1944. The disinfection and maintenance of trout hatcheries for the control of disease, with special reference to furunculosis. Amer. Fish. Soc., Trans., 74:26–34.

Ordal, E. J., and R. R. Rucker. 1944. Pathogenic myxobacteria. Soc. Exper. Biol. and Med., Proc., 56:15–18.

Phillips, Arthur M., Jr. 1956. The nutrition of trout: I. General feeding methods. Prog. Fish-Cult., 18(3):113–119.

Phillips, Arthur M., Jr., and Donald R. Brockway. 1956. The nutrition of trout: II. Protein and carbohydrate. Prog. Fish-Cult., 18(4):159–164.

————. 1957. The nutrition of trout: IV. Vitamin requirements. Prog. Fish-Cult., 19(3):119–123.

Rucker, Robert R. 1948. New compounds for the control of bacterial gill diseases. Prog. Fish-Cult., 10(1):19–22.

————. 1949. Fact and fiction in spawntaking: Addenda. Prog. Fish-Cult., 11(1):75–77.

Rucker, R. R., W. J. Whipple, J. R. Parvin, and C. A. Evans. 1953. A contagious disease of salmon possibly of virus origin. U.S. Fish and Wildl. Serv., Fishery Bull. 76, 54:35–46.

Rutter, Cloudsley. 1902. Natural history of the quinnat salmon. U.S. Fish Comm., Bull., 22:65–141.

Seaman, Wayne R. 1951. Notes on a bacterial disease of rainbow trout in a Colorado hatchery. Prog. Fish-Cult., 13(3):139–141.

Shapovalov, Leo. 1947. Distinctive characters of the species of anadromous trout and salmon found in California. Calif. Fish and Game, 33(3):185–190.

Shapovalov, Leo, William A. Dill, and Almo J. Cordone. 1959. A revised check list of the freshwater and anadromous fishes of California. Calif. Fish and Game, 45(3):159–180.

Shaw, Paul A. 1936. Hatchery trough aerators. Calif. Fish and Game, 22(2):126–136.

————. 1946. Oxygen consumption of trout and salmon. Calif. Fish and Game, 32(1):3–12.

Simon, James R., and Floyd Roberts. 1941. A drip incubator and heater combination. Prog. Fish-Cult., (56):10–13.

Skinner, John E. 1955. Use of Dow Corning antifoam AF emulsion to prevent foaming in fish transport tanks. Calif. Dept. Fish and Game, Inland Fish. Branch, Inform. Leaflet, (3):1–2. (Mimeo.)

Slater, Daniel W. 1948. Experiment on the control of columnaris with sulfa drugs. Prog. Fish-Cult., 10(3):141–142.

Snieszko, S. F. 1952. Ulcer disease in brook trout *(Salvelinus fontinalis)*: Its economic importance, diagnosis, treatment, and prevention. Prog. Fish-Cult., 14(2):43–49.

————. 1953. Virus diseases in fishes: Outlook for their treatment and prevention. Prog. Fish-Cult., 15(2):72–74.

————. 1957. Disease resistant and susceptible populations of brook trout *(Salvelinus fontinalis)*. *In:* Contributions to the study of subpopulations of fishes. U.S. Fish and Wildl. Serv., Spec. Sci. Rept., Fish., (208):126–128.

Snieszko, S. F., and G. L. Bullock. 1957. Determination of the susceptibility of *Aeromonas salmonicida* to sulfonamides and antibiotics, with a summary report on the treatment and prevention of furunculosis. Prog. Fish-Cult., 19(3):99–107.

Snieszko, S. F., and S. B. Friddle. 1950. A contribution to the etiology of ulcer disease of trout. Amer. Fish. Soc., Trans., 78:56–63.

Snieszko, S. F., and Philip J. Griffin. 1955. Kidney disease in brook trout and its treatment. Prog. Fish-Cult., 17(1):3–13.

Snieszko, S. F., and E. M. Wood. 1954. The effect of some sulfonamides on the growth of brook trout, brown trout, and rainbow trout. Amer. Fish. Soc., Trans., 84:86–92.

Snieszko, S. F., P. J. Griffin, and S. B. Friddle. 1952. Antibiotic treatment of ulcer disease and furunculosis in trout. Seventeenth No. Amer. Wildl. Conf., Trans., 17:197–213.

Snieszko, S. F., E. M. Wood, and W. T. Yasutake. 1957. Infectious pancreatic necrosis in trout. A. M. A. Archives of Pathology, 63:229–233.

Taft, A. C. 1933. Methods for counting small fish in hatcheries. Calif. Fish and Game, 19(2):122–126.

van Duijn, C., Jr. 1956. Disease of fishes. Dorset House, London, England, 174 p.

Van Winkle, Walton, and Frederick M. Eaton. 1910. The quality of the surface waters in California. U. S. Geol. Surv., Water Supply Paper no. 237, 142 p.

von Bayer, H. 1950. A method of measuring fish eggs. Prog. Fish-Cult., 2(12):105–107.

Wagner, Edward D., and Curtis L. Perkins. 1952. *Pseudomonas hydrophila,* the cause of "red mouth" disease in rainbow trout. Prog. Fish-Cult., 14(3):127–128.

Wales, J. H. 1941. Development of steelhead trout eggs. Calif. Fish and Game, 27(4):250–260.

————. 1946. Castle Lake trout investigation. First phase: Interrelationships of four species. Calif. Fish and Game, 32(3):109–143.

————. 1957. Trout of California. Calif. Dept. Fish and Game, Pamphlet, revised, Dec., 56 p.

————. 1958. Two new blood fluke parasites of trout. Calif. Fish and Game, 44(2):125–136.

Wales, J. H., and H. Wolf. 1955. Three protozoan diseases of trout in California. Calif. Fish and Game, 41(2):183–187.

Watson, Stanley W. 1953. Virus diseases of fish. Amer. Fish. Soc., Trans., 83:331–341.

Watson, Stanley W., R. W. Guenther, and R. R. Rucker. 1954. Virus disease of sockeye salmon; interim report. U. S. Fish and Wildl. Serv., Spec. Sci. Rept., Fish., (138):1–36.

Wolf, Louis E. 1936. Structure of the egg. New York Cons. Dept., Fish-Cult., 2(11):1–5.

_____. 1938. Observations on ulcer disease of trout. Amer. Fish. Soc., Trans., 68:136–151.

_____. 1945. Dietary gill disease of trout. New York Cons. Dept., Fish. Res. Bull., (7):1–30.

_____. 1947. Sulfamerazine in the treatment of trout for furunculosis and ulcer disease. Prog. Fish-Cult., 9(3):115–124.

_____. 1951. Diet experiments with trout. Prog. Fish-Cult., 13(1):17–24.

Wood, E. M., and W. A. Dunn. 1948. Fact and fiction in spawntaking. Prog. Fish-Cult., 10(2):67–72.

Wood, E. M., and W. T. Yasutake. 1956. Histopathology of fish: III. Peduncle ("cold-water") disease. Prog. Fish-Cult., 18(2):58–61.

_____. 1957. Histopathology of fish: V. Gill disease. Prog. Fish-Cult., 19(1):7–13.

Wood, E. M., S. F. Snieszko, and W. T. Yasutake. 1955. Infectious pancreatic necrosis in brook trout. A. M. A. Archives of Pathology, 60:26–28.

Wood, E. M., W. T. Yasutake, and S. F. Snieszko. 1954. Sulfonamide toxicity in brook trout. Amer. Fish. Soc., Trans., 84:155–160.

Wood, James W., and Joe Wallis. 1955. Kidney disease in adult chinook salmon and its transmission by feeding to young chinook salmon. Fish Comm. Oregon, Res. Briefs, 6(2):32–40.

Young, O. C. 1950. Quality of fresh and frozen fish and facilities for freezing, storing and transporting fishery products. Food Technology, 4(11):447-450.

APPENDIX A.   CONVERSION TABLES

Common Water Measurements Useful to Fisheries Workers

1 acre-foot_____43,560 cubic feet
1 acre-foot_____325,850 gallons
1 cubic foot of water_____7.48 gallons
1 cubic foot of water_____62.4 pounds
1,000,000 cubic feet_____22.95 acre-feet
1 gallon_____231 cubic inches
1 gallon_____0.1337 cubic feet
1 gallon of water_____8.34 pounds
100 gallons per minute_____0.223 second-feet
100 gallons per minute_____0.442 acre-feet per day
1,000,000 gallons per day_____1.55 second-feet
1,000,000 gallons per day_____3.07 acre-feet
1,000,000 gallons per day_____694 gallons per minute
1 horsepower_____1 second-foot of water falling 8.8 feet
1 inch deep on 1 square mile_____2,323,000 cubic feet
1 inch deep on 1 square mile_____0.0735 second-feet annually
100 miner's inches *_____4.96 acre-feet per day
100 miner's inches *_____18.7 gallons per second
1 second-foot_____1 acre-inch per hour
1 second-foot_____86,400 cubic feet per day
1 second-foot_____approximately 2 acre-feet per day
1 second-foot_____7.48 gallons per second or 448.8 gallons per minute
1 second-foot_____646,317 gallons per day
1 second-foot_____40 miner's inches *

$$\frac{\text{Second-feet} \times \text{fall in feet}}{11} = \text{horsepower on waterwheel operating at 80 percent efficiency.}$$

$$\frac{\text{Acre-feet} \times 43,560}{86,400 \times X} = \text{second-feet discharge over a period } X \text{ days.}$$

Discharge in second-feet
   1 second-foot for 1 day = 1.983 acre-feet; for 30 days— 59.5 acre-feet
   2 second-feet for 1 day = 3.967 acre-feet; for 30 days—119.0 acre-feet
   3 second-feet for 1 day = 5.950 acre-feet; for 30 days—178.5 acre-feet
   4 second-feet for 1 day = 7.934 acre-feet; for 30 days—238.0 acre-feet
   5 second-feet for 1 day = 9.917 acre-feet; for 30 days—297.5 acre-feet
   6 second-feet for 1 day = 11.90 acre-feet; for 30 days—357.0 acre-feet
   7 second-feet for 1 day = 13.88 acre-feet; for 30 days—416.0 acre-feet
   8 second-feet for 1 day = 15.87 acre-feet; for 30 days—476.0 acre-feet
   9 second-feet for 1 day = 17.85 acre-feet; for 30 days—535.5 acre-feet

* The second-foot (cubic foot per second = c.f.s.) is the universal unit for determining water measurements. The miner's inch is used in sections where mining activities are common. It is not standard and in California two standards are in common use. In the northern parts of the State 1 second-foot is considered to equal 40 miner's inches, while in the southern part the second-foot is considered to equal 50 miner's inches. In other western states, other values are placed on the miner's inch, so that when miner's inches are used in water measurements it is necessary to apply the standard in use in that particular area. In the above, the miner's inch (40 to 1 second-foot) is used. See California law 1901.

Common Water Measurement
Equivalents

Unit	Gallon	Quart	Pint	Pound	Avoirdupois ounce	Fluid ounce
1 gallon	1.0	4.0	8.0	8.345	133.52	128.0
1 quart	0.25	1.0	2.0	2.086	33.38	32.0
1 pint	0.125	0.5	1.0	1.043	16.69	16.0
1 pound	0.12	0.48	0.96	1.0	16.0	15.35
1 ounce	0.0075	0.03	0.06	0.0625	1.0	0.96
1 fluid ounce	0.0078	0.031	0.062	0.062	1.04	1.0
1 cubic inch	0.0043	0.017	0.035	0.036	0.573	0.554
1 cubic foot	7.481	29.922	59.848	62.428	998.848	957.48
1 cubic centimeter	0.0003	0.001	0.002	0.002	0.035	0.034
1 liter	0.264	1.057	2.1134	2.205	35.28	33.815
1 gram			0.002	0.0022	0.0353	0.034

Unit	Cubic inch	Cubic foot	Milliliter	Liter	Gram
1 gallon	231.0	0.1337	3,785.4	3.785	3,785.4
1 quart	57.749	0.0334	946.36	0.95	946.35
1 pint	28.875	0.0167	473.18	0.47	473.18
1 pound	27.67	0.016	453.59	0.454	453.59
1 ounce	1.73	0.001	28.3	0.03	28.35
1 fluid ounce	1.8		29.57	0.03	29.41
1 cubic inch	1.0	0.0006	16.39	0.0164	16.3
1 cubic foot	1,728.0	1.0	28,322.0	28.316	28,318.58
1 milliliter	0.061		1.0	0.001	1.0
1 liter	61.025	0.0353	1,000.0	1.0	1,000.0

Water Velocity and Flow Measurements

The measurement of water velocity and volume of flow under field conditions may be determined as follows:

Velocity
1. Locate two points 100 feet apart.*
2. Time in seconds for float to drift distance between points gives number of seconds per 100 feet.
3. Compute number of feet traveled per second.

Volume (or rate) of flow

Formula: $R = \dfrac{W\,Da\,L}{T}$

Where R = volume of flow in cubic feet per second.
     W = average width of stream in feet.
     D = average depth in feet.
     a = constant factor for bottom type.
        Smooth sand, etc. = 0.9
        Rough rocks, etc. = 0.8
     L = length of stream section measured.
     T = time in seconds for float to travel the measured distance.

Hydraulic Equivalents

1. Velocity of flow in a pipe in feet per second = $\dfrac{\text{gpm.} \times 0.408}{(\text{diameter in inches})^2}$
2. Doubling the diameter of a pipe increases its capacity four times.
3. Each foot elevation of a column of water produces a pressure of about one-half pound per square inch.

---

* If it is impractical to use 100 feet as a unit for determination of velocity, any convenient distance may be used.

4. The gallons per minute a pipe will deliver equals the square of the inside diameter, multiplied by the velocity in feet per second, divided by 0.408.
5. The capacity of a pipe or cylinder in gallons equals the square of the inside diameter in inches multiplied by the length in inches and by 0.0034.
6. The discharge from any pipe in cubic feet per minute equals the square of the inside diameter multiplied by the velocity in feet per minute and by 0.00545.

Capacity of Round Tanks One Foot in Depth in U. S. Gallons *

Diameter of tank in feet	Number of gallons	Cubic feet and area in square feet	Diameter of tank in feet	Number of gallons	Cubic feet and area in square feet
1.0	5.87	0.785	11.0	710.90	95.03
1.5	13.22	1.767	11.5	776.99	103.87
2.0	23.50	3.142	12.0	845.35	113.10
2.5	36.72	4.909	12.5	918.00	122.72
3.0	52.88	7.069	13.0	992.91	132.73
3.5	71.97	9.621	13.5	1,070.80	143.14
4.0	94.00	12.566	14.0	1,151.50	153.94
4.5	118.97	15.90	14.5	1,235.30	165.13
5.0	146.88	19.63	15.0	1,321.90	176.71
5.5	177.72	23.76	15.5	1,411.50	188.69
6.0	211.51	28.27	16.0	1,504.10	201.06
6.5	248.23	33.18	16.5	1,599.50	213.82
7.0	287.88	38.48	17.0	1,697.90	226.98
7.5	330.48	44.18	17.5	1,799.30	240.53
8.0	376.01	50.27	18.0	1,903.60	254.47
8.5	424.48	56.75	18.5	2,010.20	268.80
9.0	475.89	63.62	19.0	2,120.90	283.53
9.5	530.24	70.88	19.5	2,234.00	298.65
10.0	587.52	78.54	20.0	2,350.10	314.16
10.5	640.74	86.59			

* To find the capacity of tanks larger than 20 feet in diameter, refer to the table for a tank of half the given size and multiply its capacity by four.

Weights (in Grams) of Chemicals Required to Produce Desired Dilutions in Known Volumes of Water *

| Dilution | U.S. gallons of water | | | | | |
	1	5	10	15	20	25
1:1,000	3.78	18.93	37.85	56.78	75.70	94.60
1:2,000	1.89	9.46	18.93	28.39	37.85	47.30
1:3,000	1.26	6.31	12.62	18.92	25.23	31.50
1:4,000	0.95	4.73	9.46	14.19	18.92	23.60
1:5,000	0.76	3.79	7.57	11.35	15.14	18.90
1:10,000	0.38	1.89	3.79	5.68	7.57	9.40
1:15,000	0.25	1.26	2.52	3.78	5.05	6.30
1:20,000	0.19	0.95	1.89	2.84	3.78	4.70
1:100,000	0.038	0.19	0.38	0.57	0.76	0.90

* The table can be used to obtain any dilution for any volume. For example, if a 1:40,000 dilution is desired in 50 gallons of water, use the figure for the 1:20,000 dilution for 25 gallons (4.70).

Concentrations

1.0 percent salt solution = 0.622 pounds salt (9.9 ounces) to 1 cubic foot water.
2.5 percent salt solution = 1.5 pounds salt (24.8 ounces) to 1 cubic foot water.
3.0 percent salt solution = 1.86 pounds salt (29.8 ounces) to 1 cubic foot water, or 0.25 pounds salt per gallon of water.
1 ppm. (part per million) = 8.34 pounds per million gallons water
1 ppm. = 0.0584 grains per gallon
1 grain per gallon = 17.12 ppm.
1 grain per gallon = 142.9 pounds per million gallons
1 pound per million gallons = 0.1199 ppm.
Oxygen in ppm. × 0.7 = cubic centimeters or milliliters per liter

Oxygen in cubic centimeters or milliliters per liter $\times$ 1.429 = oxygen in ppm.
Carbon dioxide in ppm. $\times$ 0.509 = cubic centimeters or milliliters per liter
Carbon dioxide in cubic centimeters or milliliters per liter $\times$ 1.964 = carbon dioxide (CO_2) in ppm.

Dosage Calculations

For a 1 percent solution, add:

38 grams per gallon
1.3 ounces per gallon
10 grams per 1,000 milliliters
38 milliliters per gallon
10 milliliters per 1000 milliliters (1 liter)

For other percent solutions, multiply by factors concerned. Thus, a 5 percent solution is 5 $\times$ 38 grams per gallon.

To prepare a 1:1,000 solution, add:

3.8 grams per gallon
1 gram per 1,000 milliliters
0.13 ounce per gallon
1 milliliter per 1,000 milliliters

For other solutions, multiply or divide by factors concerned:

1:50,000 is 1/50 as strong as a 1:1,000 solution. Divide 3.8 grams by 50 and add this amount per gallon.
A 1:500 solution is twice as strong as 1:1,000. Multiply 3.8 grams by 2 to prepare this solution.

Feeding Drugs

To feed a 1 percent level in food, add:

4.5 grams per pound of food
0.2 ounces per pound of food
83 grains per pound of food

For other dosage levels, multiply by appropriate factor.

Example: For a 2 percent diet level, multiply 4.5 grams by 2 and add to 1 pound of food.
For a 0.2 percent level, multiply by 0.2.

Disinfecting Solutions

For distribution units, troughs, tanks, etc.

Active agent—Chlorine, in sodium hypochlorite
Chlorox—1 quart to 83 gallons of water
1 pint to 42 gallons of water
½ pint to 21 gallons of water
Hilex—1 quart to 55 gallons of water
1 pint to 27 gallons of water
½ pint to 13.5 gallons of water

Conversion Table, Grams per Liter to Ounces per Gallon

Grams per liter	Ounces (avoir.) per gallon (U.S.)	Percent solution	Grams per liter	Ounces (avoir.) per gallon (U.S.)	Percent solution
100	13.35	10	9	1.20	0.9
90	12.02	9	8	1.07	0.8
80	10.68	8	7	0.93	0.7
75	10.01	7.5	6	0.80	0.6
70	9.35	7	5	0.67	0.5
60	8.01	6	4	0.53	0.4
50	6.68	5	3	0.40	0.3
40	5.34	4	2	0.27	0.2
30	4.01	3	1	0.13	0.1
25	3.34	2.5	0.75	0.10	0.075
20	2.67	2	0.5	0.07	0.05
10	1.34	1	0.25	0.03	0.025

Conversion Table, Temperature in Degrees Fahrenheit and Centigrade

Degrees C.	Degrees F.	Degrees C.	Degrees F.
0	32.0	52	125.6
2	35.6	54	129.2
4	39.2	56	132.8
6	42.8	58	136.4
8	46.4	60	140.0
10	50.0	62	143.6
12	53.6	64	147.2
14	57.2	66	150.8
16	60.8	68	154.4
18	64.4	70	158.0
20	68.0	72	161.6
22	71.6	74	165.2
24	75.2	76	168.8
26	78.8	78	172.4
28	82.4	80	176.0
30	86.0	82	179.6
32	89.6	84	183.2
34	93.2	86	186.8
36	96.8	88	190.4
38	100.4	90	194.0
40	104.0	92	197.6
42	107.6	94	201.2
44	111.2	96	204.8
46	114.8	98	208.4
48	118.4	100	212.0
50	122.0		

Conversion Formulas

Degrees Centigrade $= 5/9 \times (°\text{ Fahrenheit } -32)$
Degrees Fahrenheit $= (9/5 \times °\text{ Centigrade}) + 32$

Conversion Tables, Inches to Millimeters

Inches	Mm.	Inches	Mm.	Inches	Mm.	Inches	Mm.
1/8	3	3 1/8	79	6 1/8	156	9 1/8	232
1/4	6	3 1/4	83	6 1/4	159	9 1/4	235
3/8	10	3 3/8	86	6 3/8	162	9 3/8	238
1/2	13	3 1/2	89	6 1/2	165	9 1/2	241
5/8	16	3 5/8	92	6 5/8	168	9 5/8	244
3/4	19	3 3/4	95	6 3/4	171	9 3/4	248
7/8	22	3 7/8	98	6 7/8	175	9 7/8	251
1	25	4	102	7	178	10	254
1 1/8	29	4 1/8	105	7 1/8	181	10 1/8	257
1 1/4	32	4 1/4	108	7 1/4	184	10 1/4	260
1 3/8	35	4 3/8	111	7 3/8	187	10 3/8	264
1 1/2	38	4 1/2	114	7 1/2	191	10 1/2	267
1 5/8	41	4 5/8	117	7 5/8	194	10 5/8	270
1 3/4	44	4 3/4	121	7 3/4	197	10 3/4	273
1 7/8	48	4 7/8	124	7 7/8	200	10 7/8	276
2	51	5	127	8	203	11	279
2 1/8	54	5 1/8	130	8 1/8	206	11 1/8	283
2 1/4	57	5 1/4	133	8 1/4	210	11 1/4	286
2 3/8	60	5 3/8	137	8 3/8	213	11 3/8	289
2 1/2	64	5 1/2	140	8 1/2	216	11 1/2	292
2 5/8	67	5 5/8	143	8 5/8	219	11 5/8	295
2 3/4	70	5 3/4	146	8 3/4	222	11 3/4	298
2 7/8	73	5 7/8	149	8 7/8	225	11 7/8	302
3	76	6	152	9	229	12	305

Conversion Table, Miscellaneous Measures and Weights

1 acre _____ 0.404687 hectare
1 acre _____ 43,560 square feet
1 acre _____ 208.71 feet square
1 centimeter _____ 0.3937 inch
1 yard _____ 91.44 centimeters
1 meter _____ 39.37 inches
1 inch _____ 2.54 centimeters
1 gram _____ 15.432 grains
1 kilogram _____ 2.205 pounds
1 pound _____ 7,000 grains

APPENDIX B.   GLOSSARY

Abdomen
> Belly; the ventral side of the fish surrounding the cavity containing the digestive and reproductive organs.

Abdominal
> Pertaining to the belly; said of the pelvic fins of fishes when inserted behind the pectorals.

Acclimatization
> The adaptation of fishes to a new environment or habitat or to different climatic conditions.

Adipose fin
> A fleshy fin-like projection without rays, behind the rayed dorsal fin.

Agglutination
> The formation of clumps or floccules by pollen, bacteria, erythrocytes, spermatozoa, and some protozoans.

Air bladder or swim bladder
> A membranous sac filled with gas situated in the body cavity of fishes, ventral to the vertebral column.

Albino
> Abnormal congenital deficiency of black pigmentation. Albinos have pink eyes because of lack of pigmentation.

Alevin
> A young fish, especially a newly-hatched salmon or trout before absorption of the yolk sac.

Alimentary tract
> The digestive tract or canal.

Ammocoete
> The larval form of lampreys.

Amphibious
> Capable of living both on land and in water.

Anadromous
> Said of fishes which migrate from salt to fresh water to spawn.

Anal
> Pertaining to the anus or vent.

Anal fin
> The fin on the ventral median line behind the anus.

Anal papilla
> A protuberance in front of the genital pore and behind the vent in certain groups of fishes.

Annulus
> A yearly mark formed by a zone of irregularities in the sculpturing on scales, corresponding to a period of slow growth.

Anterior
> In front of, or toward the head end.

Anus
> The external posterior opening of the intestine; the vent.

Aquarium (pl. aquaria)
> A tank or other suitable container in which fishes and other aquatic organisms may be maintained.

Artery
> A blood vessel carrying blood away from the heart.

Articulate
> Jointed; said of the structure of soft fin rays.

Assimilation
> The transformation of digested nutriments into the fluids of an organism by a process of constructive metabolism.

Atrium
> The auricular portion of the heart.

Atrophy
> Nondevelopment; diminution in size.

Attenuate
> Long and slender.

Auditory
> Referring to the ear or to hearing.

Axilla
> The region just behind or under the pectoral fin base.

Backbone
> Vertebral column.

Bacterium (pl. bacteria)
> One of a large, widely distributed group of typically one-celled microorganisms, often parasitic.

Bar
> Vertical color mark on fishes.

Barbel
> An elongated fleshy projection, usually about the head.

Barbiturate
> One of a large group of drugs used as sedatives, hypnotics, etc.

Basal
> Pertaining to the base; at or near the base of a fin.

Bicuspid
> Having two points, split into two parts.

Blastoderm
> The foundation from which the embryo will form. For practical purposes, the blastoderm is the same as the blastodisc or germinal disc.

Blastodisc
> See blastoderm.

Blastopore
> As the blastoderm grows over the egg it finally leaves a circular opening or blastopore.

Blastula
> A hollow ball of cells, one of the early stages in embryological development.

Body
> The region from the gill openings to the anus.

Bony fishes
> Fishes having a hard calcified skeleton as contrasted with a cartilaginous one.

Branchiae
> Gills, the respiratory organs of fishes.

Branchiocranium
> The bony skeleton supporting the gill arches.

Branchiostegals
> The bony rays supporting the branchiostegal membrane, under the head of fishes, below the opercular bones, behind the lower jaw, and attached to the hyoid arch.

Breast
> An area with indefinite boundaries between the pelvic fins and the isthmus; sometimes called chest.

Buccal
> Pertaining to the mouth.

Buccal disc
> A circular funnel-like structure around the mouth in lampreys.

Buccal incubation
> Incubation of eggs in the mouth; oral incubation.

Bulbus arteriosus

> The blood-collecting chamber between the heart and the ventral aortae. The bulbus arteriosus is not a contracting chamber, hence is actually part of the following artery system rather than of the heart itself.

Caecal

> Of the form of a blind sac.

Caecum (pl. caeca)

> An appendage in the form of a blind sac, connected with the alimentary canal, such as one of the pyloric caeca at the posterior end of the stomach or pylorus.

Canine teeth

> Elongated, conical teeth on the jaws, much longer than the other teeth.

Carnivorous

> Eating flesh; feeding or preying on animals.

Cartilage

> A substance more flexible than bone but serving the same purpose. A trout's skeleton is at first all cartilage but later part of it changes to bone.

Caseous

> Cheesy.

Catadromous

> Said of fishes which migrate from fresh to salt water to spawn.

Catalyst

> An agent, e.g., an enzyme, which can accelerate or retard, or initiate, a reaction and apparently remains unchanged.

Caudal

> Pertaining to the tail.

Caudal fin

> The unpaired fin at the posterior end of the body; the tail fin of fishes.

Caudal peduncle

> The tapering or slender portion of the body behind the base of the last ray of the anal fin.

Cell

> A microscopic unit in the structure of the body of a fish (and other living organisms). Each cell is surrounded by a membrane and has a nucleus and other distinctive features, but in one part of the body the cells may be quite different from those in another part. For example, the cells in the skin of a fish look much different from those in the liver.

Centrum

> The body of a vertebra.

Cercaria (pl. cercariae)

> A heart-shaped trematode larva with a tail.

Cerebellum

> A single lobe of the brain situated at the top and rear.

Cerebral hemispheres

> The front lobes of the brain.

Cheek

> The fleshy area behind and below the eye and anterior to the opercle.

Chin

> The region at the tip of the lower jaw.

Chlorophyll

> The green coloring matter found in plants and in some animals.

Chromatophores

> Colored pigment cells.

Chromosomes

> Units of heredity in the nucleus of cells.

Circuli

> The more or less concentric growth marks in a fish scale.

Class
> In classification, a division of a phylum, divided into orders.

Cloaca
> The common cavity into which the rectal, urinary, and genital ducts open.

Coelomic cavity
> The body cavity containing the internal organs.

Colloid
> A gelatinous substance which does not readily diffuse through an animal or vegetable membrane.

Compressed
> Flattened from side to side, as in the case of a sunfish.

Cornea
> Outer covering of the eye.

Cranium
> The part of the skull enclosing the brain.

Ctenoid scales
> Scales with minute spines on their distal exposed portions. The spines can be felt by gently rubbing the fish with the finger, or they can be seen with a lens.

Cycloid scales
> Scales without spines, but with concentric lines called circuli and annuli. Such scales are smooth to the touch.

Cyst
> The capsule or enclosing membrane round certain cells, such as bacteria and protozoans in the resting stage.

Cytoplasm
> The contents of a cell, exclusive of the nucleus.

Deciduous
> Temporary, falling off easily; said of the scales of certain fishes.

Degossypolize
> To remove the toxic compound gossypol from cottonseed.

Dentary bones
> The principal or anterior bones of the lower jaw or mandibles. They usually bear teeth.

Depressed
> Flattened in the up and down direction.

Depth of fish
> The greatest vertical diameter; usually taken just in front of the dorsal fin.

Dermal
> Pertaining to the skin.

Dermis
> The skin.

Distal
> The remote or extreme end of a structure.

Dorsal
> Pertaining to the back.

Dorsal fin
> The fin on the back or dorsal side, in front of the adipose fin if it is present.

Dorsal fin ray
> One of the cartilaginous rods which support the membranes of the fin on the back of a fish.

Ductus pneumaticus
> A tube that connects the air bladder with the esophagus. (Absent in black bass, sunfishes, and other spiny-rayed fishes.)

Ectoderm
> The outer layer of cells which gives rise to various organs as the embryo develops.

Ectoparasite
> A parasite living outside the body of a fish.

Egg
> An ovum which when fertilized may develop into an animal.

Electrolyte
> A substance in which the conduction of electricity is accompanied by chemical decomposition.

Emarginate fin
> Fin with the margin containing a shallow notch, as in the caudal fin of the rock bass.

Embryo
> A young organism in the early stages of development, before it becomes self-supporting.

Endocrine
> A ductless gland.

Endoparasite
> A parasite living inside a fish.

Endoskeleton
> The skeleton proper; the inner bony framework.

Enzyme
> A catalyst produced by living organisms and acting on one or more specific substrates.

Epidermis
> The outer layer of the skin.

Esophagus
> The gullet, a musculomembranous canal extending from the pharynx to the stomach.

Excretion
> The process of getting rid of or throwing off waste products by any organism.

Exudation
> Any discharge through an incision or pore.

Exoskeleton
> The hard bony parts on the exterior surfaces, such as scales, scutes, and bony plates.

Fauna
> The animals inhabiting any region, taken collectively.

Fertilization
> The union of the sperm and egg.

Finfold
> A ridge of tissue on the embryo which gives rise to one or more of the fins.

Fish
> Any of numerous, cold-blooded, aquatic, water-breathing, craniate vertebrates having the limbs developed as fins. When used in the plural, fish refers to two or more specimens of the same species.

Fishes
> The plural of fish used when referring to two or more kinds of fish.

Flagellated
> Furnished with flagella.

Flagellum
> A lash-like process, as in some Protozoa and cells.

Flash
> A term used to describe the quick turning movements of fish, especially when annoyed by external parasites, causing a momentary reflection of light from their sides and bellies. In flashing, fish often scrape the sides and bottom of the pond to rid themselves of the parasites.

Foramen
> A hole or opening.

Fork length
> The distance from the tip of the snout to the fork of the caudal fin.

Frenum
> A small piece of skin binding the lip to the edge of the jaw.

Fry
> The young of fishes.

Fungus
> Any of a group of thallophytic plants (Fungi), lacking chlorophyll. Fungi comprise the molds, mildews, rusts, smuts, mushrooms, and others; some kinds are parasitic on fishes.

Gall bladder
> The body vessel containing bile.

Gape
> The opening of the mouth.

Gastrula
> The embryonic stage of development consisting of two layers enclosing a sac-like central cavity with a pore at one end.

Genetics
> The science of heredity and variation.

Genital
> Pertaining to the region of the reproductive organs.

Genus
> One of the subdivisions of the family, consisting of one or a group of closely related species.

Germinal disc
> The disc-like area of an egg yolk on which segmentation first appears; blastodisc.

Gill arch
> The cartilage which supports the gill filaments. The trout has four on each side of its head.

Gill bud
> The embryonic stage of the future gill.

Gill clefts or slits
> Spaces between the gills connecting the pharyngeal cavity with the gill chamber.

Gill cover
> The flap-like cover of the gill and gill chamber; the opercle.

Gill filament
> The slender, delicate, fringe-like structure composing the gill.

Gill membranes
> The skin or dermal membranes supported by the branchiostegals, more or less restricting the gill opening in the region of the isthmus.

Gill openings
> The external openings of the gill chambers. A single pair is present in all true fishes found in fresh water.

Gill rakers
> A series of bony appendages, variously arranged along the anterior and often the posterior edges of the gill arches.

Gills
> The highly vascular, fleshy filaments used in aquatic respiration.

Globulin
> One of a group of proteins insoluble in water, but soluble in dilute solutions of neutral salts.

Gonads
> The reproductive organs; testes and ovaries.

Guanin
> A waste product of the blood with the power of reflecting light.

Gullet
> The esophagus.

Hermaphrodite
> An individual that contains the reproductive or generative organs of both sexes.

Heterocercal
> Said of the tail of fishes when the upper and lower lobes are not equal in length.

Holotype
> The specimen on which the description of a new species is based.

Homocercal

Said of the tail of fishes when the lobes are equal in length, as in an adult trout, or when the tail is pointed.

Hormone

A substance normally produced in certain cells of the body, transported to other distant cells, and necessary for the proper functioning of the body; an internal secretion of ductless glands which passes into blood vessels by osmosis.

Hybrid

The offspring from crossing two different species

Hyoid bones

Bones in the floor of the mouth, supporting the tongue.

Hypurals

The modified plate-like last few vertebrae supporting the caudal fin rays.

Ichthyophthirius ("Ich")

A protozoan parasite of fishes producing "white-spot disease".

Ichthyology

The science of the study of fishes.

-id (suffix)

Indicating membership in a family, thus salmonid, a member of the Salmonidae.

-idae (suffix)

The family name always ends in -idae, as in Salmonidae.

Imbricate

Overlapping, like the shingles on a roof.

Inarticulate

Not jointed.

Incisors

Teeth compressed to form a chisel-like cutting edge.

Incubation

The process of development of embryos to the hatching period.

Inferior mouth

Mouth decidedly on the under side of the head, opening downward.

Infraoral

Below the mouth. The teeth of the mouth or disc in lampreys below the oral opening.

Inner ear

The auditory organ of vertebrates.

insertion of fin

A term applied to the point where the paired fins arise from the body.

Interbranchial septum

The membrane separating the two gill filaments on a gill arch.

International Commission on Zoological Nomenclature

A group of men who form and interpret the rules of zoological nomenclature.

Interorbital space

The space between the eyes on the dorsal side of the head. The least width of the bony interorbital space is measured unless the fleshy interorbital space is indicated.

Interspinals

Bones to which the rays of the fins are attached.

Intestine

The lower part of the alimentary canal from the pyloric end of the stomach to the anus.

Invaginate

To fold in, as when a rubber ball is pushed in at one point.

Iris

The curtain stretched across the aqueous chamber of the eye, in front of the lens, having a contractile aperture called the pupil.

isocercal
>Said of the tail of fishes when the last vertebrae progressively become smaller and smaller and end in the median line of the caudal fin, the hypural plate being nearly obsolete.

Isotonic
>Of equal tension; having equal osmotic pressure.

Isthmus
>The region just anterior to the breast of a fish where the gill membranes converge; the fleshy interspace between gill openings.

Kidney
>One of the pair of glandular organs in the abdominal cavity which serve to excrete urine.

Lacustrine
>Living in lakes.

Larva (pl. larvae)
>An immature form, which must undergo change of appearance or pass through a metamorphic stage to reach the adult state.

Lateral
>Pertaining to the side.

Lateral band
>A horizontal pigmented band along the sides of a fish.

Lateral line
>A series of sensory tubes or pores opening to the exterior, sometimes through scales or a sensory canal, along the sides of a fish.

Length
>The term length may refer to the total length, fork length, or standard length (see under each item).

Lentic
>Pertaining to standing water; lenitic.

Leucocyte
>A white blood corpuscle.

Lingual
>Pertaining to the tongue.

Littoral
>pertaining to the shore.

Liver
>The glandular organ which secretes bile.

Lotic
>Pertaining to running water.

Malpighian body
>The functional unit of the kidney, consisting of Bowman's capsule, the glomerulus, and the uriniferous tubule.

Mandible
>Lower jaw.

Maxilla or maxillary
>The hindmost bone of the upper jaw.

Melanophore
>A black pigment cell; large numbers of these give trout their dark color.

Membrane
>A thin "skin".

Mesentery
>A fold of the peritoneum that invests the intestine and supports it from the body wall.

Metamorphosis
>The rapid change in anatomical structure that transforms the larva or postlarva into the adult.

Micropyle
> Aperature in the egg membrane for admission of a spermatozoon.

Miliary
> Of granular appearance; consisting of small and numerous grain-like parts.

Milt
> The sperm of fishes.

Miracidium
> The ciliated embryo or youngest stage of a trematode.

Mottled
> Blotched; color spots running together.

Mucus
> A viscid or slimy substance secreted by the mucous glands.

Myomere
> An embryonic muscular segment which later becomes a section of the side muscle of a fish.

Myotome
> Muscle segment.

Nape
> Region just behind the occiput.

Necrosis
> Death of tissue.

Neural arch
> The dorsal arch of a vertebra for the passage of the spinal cord.

Neural canal
> The cavities formed by the neural arches as a whole.

Neural spine
> The uppermost spine of a vertebra.

Nomenclature
> System of naming animals, plants, organs, etc.

Nostril
> An opening of the nasal chamber.

Notochord
> The embryonic rod around which the vertebral column forms.

Obtuse
> Blunt.

Occiput
> The posterior dorsal part of the head or skull.

Opercle
> Gill cover.

Opercular bone
> The flat, more or less triangular bone supporting the gill cover or opercle.

Opercular flap
> The fleshy prolongation of the upper posterior angle of the opercle.

Operculum
> Gill cover or opercle.

Optic
> Referring to the eye; e.g., the optic lobes of the brain, which are connected to and control the eyes.

Orbit
> The cavity of the skull containing the eye.

Order
> In classification, a group of allied organisms ranking between family and class.

Osmosis
> The diffusion which takes place between two fluids or solutions through a permeable or semipermeable membrane, and tending to equalize their concentrations.

Osseous
>   Bony.

Otic
>   Pertaining to the ear.

Otoliths
>   Two or three small bones, somewhat spherical in shape, situated in the sacculus of the inner ear of fishes.

Ovaries
>   The female reproductive organs which give rise to the eggs.

Oviduct
>   The tube which carries eggs from the ovary to the exterior.

Oviparous
>   Producing eggs which are fertilized, develop, and hatch after expulsion from the body.

Ovoviviparous
>   Producing eggs, usually with much yolk, which are fertilized internally. Little or no nourishment is furnished by the mother during development; hatching may occur before or after expulsion.

Ovum (pl. ova)
>   Egg.

Paired fins
>   The pectoral and pelvic fins.

Palate
>   The roof of the mouth.

Palatines
>   Bones just back of the vomer in the roof of the mouth, one on each side.

Papilla (pl. papillae)
>   A small fleshy projection.

Papillose
>   Covered with papillae.

Paratype
>   An additional specimen, other than the holotype, on which the description of a new species is based.

Parr mark
>   One of the vertical color bars found on salmonids and certain other fishes.

Pectoral fins
>   The anterior or dorsalmost paired fins in fishes, corresponding to the anterior limbs of the higher vertebrates.

Pectoral girdle
>   The bones supporting the pectoral fins.

Pelvic fins
>   Paired fins corresponding to the posterior limbs of the higher vertebrates (sometimes called ventral fins).

Pelvic girdle
>   The bones supporting the pelvic fins.

Perforate
>   Pierced through.

Peritoneum
>   The membrane lining the abdominal cavity.

Perivitelline
>   Surrounding the yolk of an egg.

Pharynx
>   The cavity behind and communicating with the mouth.

Photosynthesis
>   The formation of carbohydrates from carbon dioxide and water which takes place in the chlorophyll-containing tissues of plants exposed to light.

Phylum
> A group of animals or plants constructed on a similar general plan; a primary division in classification.

Pigment
> Coloring matter.

Pituitary
> An endocrine gland attached to the brain.

Premaxillaries
> The bones, one on each side, forming the front of the upper jaw in fishes; usually they bear teeth.

Protozoan (pl. protozoans)
> A member of the phylum Protozoa, composed of minute, single-celled animals reproducing by fission; most are aquatic and some are parasites.

Protoplasm
> Living cell substance.

Protractile
> Capable of being drawn in or thrust forward, as the upper jaw in many fishes.

Proximal
> Nearest; basal.

Pseudocyst
> A residual protoplasmic mass which swells and ruptures, liberating spores.

Pterygiophore
> One of the supporting bones for each ray of the dorsal and anal fins.

Pupil of eye
> The blackish central part of the eye surrounded by the iris.

Purulent
> Of or consisting of pus.

Pyloric caeca
> *See* caecum.

Radii of scale
> Lines on the proximal part of a scale, radiating from near center to base.

Ray
> A supporting rod for a fin. There are two kinds: hard (spines) and soft rays.

Red gland
> A gland composed of a large group of capillaries situated on the air bladder.

Reticulate
> Marked with a network of lines.

Roe
> The eggs of fishes.

Rostrum
> The snout.

Rudimentary
> Undeveloped or nearly so.

Sacculus
> A small sac of the internal ear.

Salinometer
> An instrument for measuring the amount of salt in a solution.

Scale formula
> A conventional formula used in identifying fishes. "Scales 7 + 65 + 12", for example, indicates 7 scales above the lateral line, 65 in the lateral line, and 12 below it.

Scales above the lateral line
> The number of scales is counted in various ways, depending on the kind of fish. Usually the count is made in an oblique row beginning with the first scale above the lateral line and running anteriorly to the base of the dorsal fin.

Scales below the lateral line
> The number of scales is counted in a row beginning at the origin of the anal fin and running obliquely dorsally either forward or backward, to the lateral line. In certain fishes this count is made from the base of the pelvic fin.

Scales in the lateral line
> Usually the number of scales bearing tubes in the lateral line is counted, or the number of oblique rows crossing the lateral line is counted just above it. The first scale counted is the one at the upper edge of the opercular opening; the last one counted is at the end of the hypural plate or base of the caudal fin rays. Scales on the caudal fin base and caudal fin, even though they possess tubes, are not included in the count.

Scaly appendage
> An accessory scale or fleshy triangular projection at the dorsal edge of the pelvic base on certain fishes.

Seasoned water
> Water which has been conditioned for aquarium fishes. Raw tap water may be conditioned by storage for seven days in a shallow, nonmetallic pan. Artificial aeration may hasten this aging.

Second dorsal
> The posterior of two fins, usually the soft-rayed dorsal fin of spiny-rayed fishes.

Septum (pl. septa)
> A thin partition.

Serous
> Watery; like serum.

Serum
> The watery portion of an animal fluid remaining after coagulation.

Sessile
> Permanently attached.

Snout
> The portion of the head in front of the eyes. The snout is measured from its most anterior tip to the anterior margin of the orbit.

Soft dorsal
> The part of the dorsal fin, sometimes all of it, composed of soft or articulated rays.

Soft fins
> Fins with soft rays only, designated as soft dorsal, etc.

Soft rays
> Fin rays that are cross-striated or articulated, like a bamboo fish pole.

Somite
> A segment of the embryo's body which will later form one of the side muscles.
> A somite is an early stage in the formation of a myomere.

Species
> A group of interbreeding individuals not ordinarily interbreeding with another such group; a systematic unit including geographic races and varieties, and included in a genus.

Spermatozoon
> A male reproductive cell, consisting usually of head, middle piece, and locomotory flagellum.

Spinal cord
> The cylindrical structure within the spinal canal, a part of the central nervous system.

Spines
> Unsegmented rays, commonly hard and pointed.

Spinous dorsal
> Anterior part of the dorsal fin of spiny-rayed fishes; any dorsal fin composed of inarticulated rays.

Spiny rays
> Pungent or non-cross-striated fin rays.

Spleen
> The organ in which lymphocytes are produced and red blood corpuscles destroyed, in vertebrates.

Standard length
> The distance from the tip of the snout to the base of the caudal fin rays.

Superior
> As applied to the mouth, the term means opening in a more dorsal or upward as opposed to anteriorly facing or ventral direction.

Syndrome
> A group of concomitant symptoms.

Synonymy
> A list of the scientific names which have been applied to the same species or other group, other than the valid name.

Tactile
> Pertaining to the sense of touch.

Tail
> The part of the body behind the body cavity.

Terminal mouth
> The mouth is so designated when situated in the horizontal axis of the head, with neither chin nor snout projecting.

Testes
> The male reproductive organs which give rise to sperm.

Titration
> A method, or the process, of determining the strength of a solution, or the concentration of a substance in solution, in terms of the smallest amount of it required to bring about a given effect in reaction with another known solution or substance.

Total length
> The distance from the tip of the snout to the tip of the caudal fin.

Transverse
> Crosswise.

Trematode
> Any of a class (Trematoda) of flatworms includiing the flukes and related organisms.

Truncate
> Abrupt, as if cut off squarely.

Type
> *See* holotype and paratype.

Type locality
> The locality or localities from which the holotype and paratypes were collected.

Urinary bladder
> The bladder attached caudally to the kidneys; the kidneys drain into it.

Urostyle
> The rudimentary or embryonic rear tip of the vertebral column which occurs on the dorsal edge of the hypural plate.

Vas deferentia
> The ducts by which sperm is conveyed to the seminal vesicles.

Vein
> A tubular vessel that carries blood to the heart.

Vent
> The external posterior opening of the alimentary canal; the anus.

Ventral aortae
> Large paired arteries which carry blood from the bulbus arteriosus to the gills.

Ventral fins
> Pelvic fins.

Vertebrae
>The bones of the spinal column.

Vertebral column
>The backbone or spinal column, composed of a series of vertebrae.

Vertical fins
>Fins along the median line of the body: the dorsal, anal, and caudal fins.

Vesicle
>A pouch or sac. For example, an optic vesicle in an embryonic eye.

Vestigial
>Rudimentary.

Villiform
>Said of the teeth of fishes when slender and crowded into velvety bands or compact patches.

Viscous
>Slimy.

Vitelline membrane
>The outer membrane of an egg.

Viviparous
>Bringing forth living young; the mother contributes food toward the development of the embryos.

Vomer
>Bone of the anterior part of the roof of the mouth, commonly triangular and often with teeth.

Xanthophore
>A yellow chromatophore.

Yolk
>The food part of an egg.

Zoogeography
>The science of the geographical distribution of animals.

Photoelectronic composition by
CALIFORNIA OFFICE OF STATE PRINTING